企业级卓越人才培养解决方案"十三五"规划教材

Angular 项目实战

天津滨海迅腾科技集团有限公司　主编

南开大学出版社
天　津

图书在版编目(CIP)数据

Angular项目实战/天津滨海迅腾科技集团有限公司主编. —天津：南开大学出版社，2018.7(2024.1重印)
ISBN 978-7-310-05615-6

Ⅰ. ①A… Ⅱ. ①天… Ⅲ. ①超文本标记语言—程序设计 Ⅳ. ①TP312.8

中国版本图书馆CIP数据核字(2018)第131973号

主　编　李树真　李春阁
副主编　雷　莹　郝振波　刘　盟　李　悦

Angular项目实战
Angular XIANGMU SHIZHAN

南开大学出版社出版发行
出版人:刘文华
地址:天津市南开区卫津路94号　邮政编码:300071
营销部电话:(022)23508339　营销部传真:(022)23508542
https://nkup.nankai.edu.cn

河北文曲印刷有限公司印刷　全国各地新华书店经销
2018年7月第1版　2024年1月第3次印刷
260×185毫米　16开本　16.75印张　420千字
定价:59.00元

如遇图书印装质量问题,请与本社营销部联系调换,电话:(022)23508339

企业级卓越人才培养解决方案“十三五”规划教材
编写委员会

企业级卓越人才培养解决方案简介

企业级卓越人才培养解决方案（以下简称“解决方案”）是面向我国职业教育量身定制的应用型、技术技能型人才培养解决方案，以教育部 - 滨海迅腾科技集团产学合作协同育人项目为依托，依靠集团研发实力，联合国内职业教育领域相关政策研究机构、行业、企业、职业院校共同研究与实践的科研成果。本解决方案坚持“创新校企融合协同育人，推进校企合作模式改革”的宗旨，消化吸收德国“双元制”应用型人才培养模式，深入践行“基于工作过程”的技术技能型人才培养，设立工程实践创新培养的企业化培养解决方案。在服务国家战略，京津冀教育协同发展、中国制造 2025（工业信息化）等领域培养不同层次的技术技能人才，为推进我国实现教育现代化发挥积极作用。

该解决方案由“初、中、高级工程师”三个阶段构成，包含技术技能人才培养方案、专业教程、课程标准、数字资源包（标准课程包、企业项目包）、考评体系、认证体系、教学管理体系、就业管理体系等于一体。采用校企融合、产学融合、师资融合的模式在高校内共建大数据学院、虚拟现实技术学院、电子商务学院、艺术设计学院、互联网学院、软件学院、智慧物流学院、智能制造学院、工程师培养基地的方式，开展“卓越工程师培养计划”，开设系列“卓越工程师班”，“将企业人才需求标准、工作流程、研发项目、考评体系、一线工程师、准职业人才培养体系、企业管理体系引进课堂”，充分发挥校企双方特长，推动校企、校际合作，促进区域优质资源共建共享，实现卓越人才培养目标，达到企业人才培养及招录的标准。本解决方案已在全国近几十所高校开始实施，目前已形成企业、高校、学生三方共赢格局。未来三年将在 100 所以上高校实施，实现每年培养学生规模达到五万人以上。

天津滨海迅腾科技集团有限公司创建于 2008 年，是以 IT 产业为主导的高科技企业集团。集团业务范围已覆盖信息化集成、软件研发、职业教育、电子商务、互联网服务、生物科技、健康产业、日化产业等。集团以产业为背景，与高校共同开展产教融合、校企合作，培养了一批批互联网行业应用型技术人才，并吸纳大批毕业生加入集团，打造了以博士、硕士、企业一线工程师为主导的科研团队。集团先后荣获：天津市“五一”劳动奖状先进集体，天津市政府授予“AAA”级劳动关系和谐企业，天津市“文明单位”，天津市“工人先锋号”，天津市“青年文明号”“功勋企业”“科技小巨人企业”“高科技型领军企业”等近百项荣誉。

前　　言

随着 Web 开发技术的不断更新，各种框架层出不穷。Google 推出了自己的 Web 开发框架，名为 Angular。Angular 的出现引起众多开发者的关注，其“开箱即用”的特点更是受到大多数人的喜爱。通过 Angular 框架即可完成大部分的前端开发工作，具有降低开发成本、快速开发等优势。解决了使用静态网页技术构建 Web 应用不足的问题，让浏览器能够显示出更多想要的效果。

本书由八个项目组成，详细讲解 Angular 项目从“创建”到“发布”的过程。项目一至项目六主要从“Angular 架构”→“Angular 数据显示”→“Angular 生命周期钩子”→“依赖注入的应用”→“Angular 路由的使用”→“HTTP 服务”等知识，通过对项目一至项目六的学习使读者能够掌握 Angular 开发所需基本知识。项目七主要介绍 Angular 项目开发后的环境测试部署。项目八详细介绍从 AngularJS 项目升级到 Angular 项目的方法。通过 Angular 开发所需基本知识以及环境测试部署，使读者掌握 Angular 项目开发的基本操作，具有独立开发 Angular 项目以及测试与部署的能力。

本书的每个项目都分为学习目标、学习路径、任务描述、任务技能、任务实施、任务总结、英语角、任务习题八个模块来讲解相应的知识点。此结构条理清晰、内容详细，任务实施可以将所学的理论知识充分的应用到实战中。本书的八个项目都与我们的生活息息相关，使读者在学习过程中加深对项目的理解，快速掌握相关技能。

本书由李树真、李春阁任主编，雷莹、郝振波、刘盟、李悦任副主编，李树真、李春阁负责统稿，雷莹、郝振波、刘盟、李悦负责整体内容的规划与编排。具体分工如下：项目一至项目三由李树真、李春阁共同编写，雷莹负责全面规划；项目四至项目六由雷莹、郝振波编写，郝振波、刘盟负责全面规划；项目七至项目八由刘盟、李悦编写，李悦负责规划。

本书理论内容简明扼要、通俗易懂、即学即用；实例操作讲解细致，步骤清晰，在本书中，操作步骤后有相对应的效果图，便于读者直观、清晰地看到操作效果，牢记书中的操作步骤。使读者在 Angular 的学习过程中能够更加顺利，为后期 Angular 的进一步学习打下坚实的基础。

天津滨海迅腾科技集团有限公司
技术研发部

目　录

项目一　初识 Angular

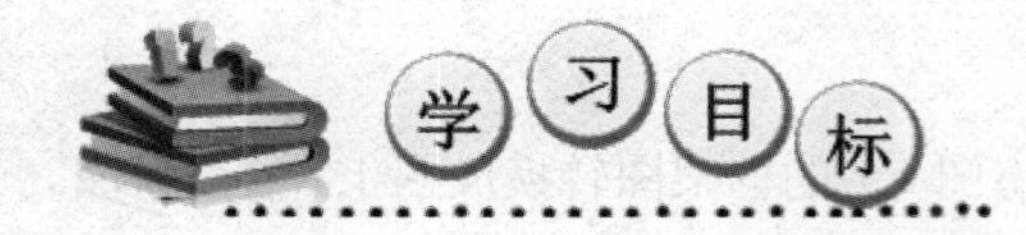

通过智慧工厂中央管理系统登录功能的实现，了解登录功能的实现流程，学习使用不同方式搭建 Angular 环境，掌握 Angular 的项目结构，具有创建 Angular 项目的能力。在任务实现过程中：

- 了解登录功能的实现流程。
- 学习 Angular 环境的搭建。
- 掌握 Angular 项目的创建。
- 具备创建 Angular 项目的能力。

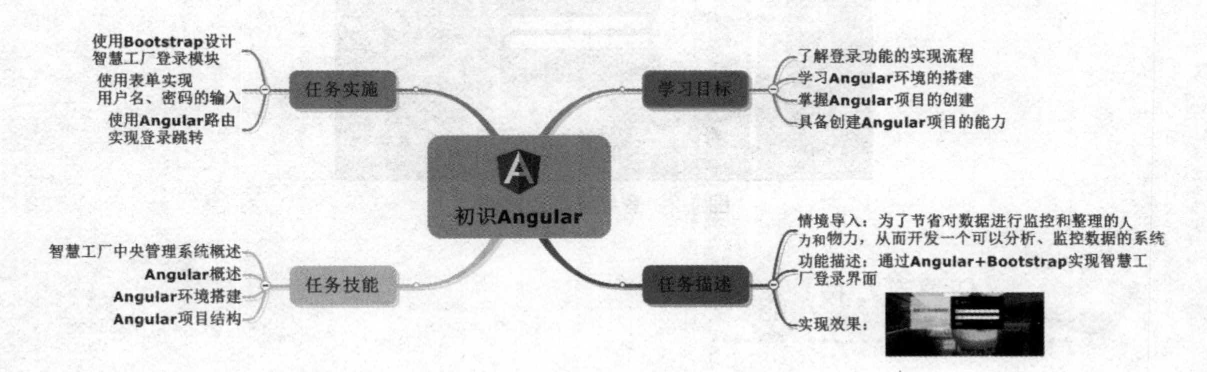

【情境导入】

在工业生产中，会产生大量的数据，工作人员需要对数据进行监控和整理，但是，通过人力来监控和整理数据会花费很多的时间与精力。因此，某公司开发了一个可以分析、监控数据的系统，此系统主要应用于工业生产中，能快速地分析、整理数据，可以提高工作效率，降低生产成本。于是该公司给此系统命名为“智慧工厂中央管理系统”，简称“智慧工厂”。本项目主要通过实现智慧工厂登录模块来学习 Angular 的环境搭建及项目创建。

【功能描述】

通过 Bootstrap+Angular 实现智慧工厂登录模块：

- 使用 Bootstrap 设计智慧工厂登录模块。
- 使用表单实现用户名、密码的输入。
- 使用 Angular 路由实现登录跳转。

【基本框架】

基本框架如图 1.1 所示，通过本项目的学习，能将图 1.1 的框架图转换成图 1.2 的效果图。

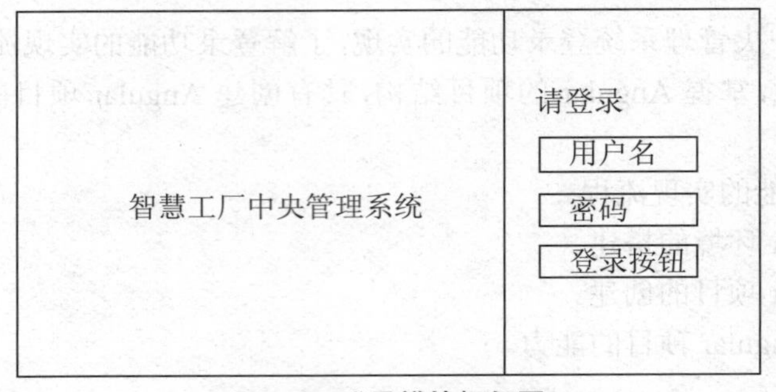

图 1.1　登录模块框架图

图 1.2　登录模块效果图

技能点 1　智慧工厂中央管理系统概述

1　智慧工厂中央管理系统项目背景

智慧工厂中央管理系统为工业化工厂生产提供了强有力的支撑，在物联网技术的基础上，采用信息传感设备进行数据采集。为了更好的实现信息化，本系统将其采集到的数据进行存储，并通过系统调用对其分析，实现数据分析和设备预警。

智慧工厂中央管理系统的提出，为工业化发展指引了新的方向，通过对设备运行状态的监控和故障监测，提高设备运行状态信息的及时性和准确性，从而满足工厂对生产流程监控的要求。智慧工厂中央管理系统能够提高工厂生产过程的可控性，减少生产线上人工的干预，及时正确地采集生产线数据，提高企业工作效率和生产能力。

2　智慧工厂中央管理系统的功能

智慧工厂中央管理系统的提出是为了更好的实现生产管理一体化、实时监控以及对预警的分析，减少人为操作可能造成的误差，为工厂的管理者提供更好的管理方式。

（1）数据采集

生产过程中产生的数据在智慧工厂中央管理系统中占据很重要的地位，因此对数据的实时性、准确性、有效性有很高的要求。智慧工厂中央管理系统采用 WinCC（数据采集与监控系统）对数据进行实时采集，并编写 WinForm 项目通过相应的查询方式读取 WinCC 数据库中数据并将其保存在服务器数据库中。图 1.3 为 WinCC 的使用机制（本书不介绍此内容）。

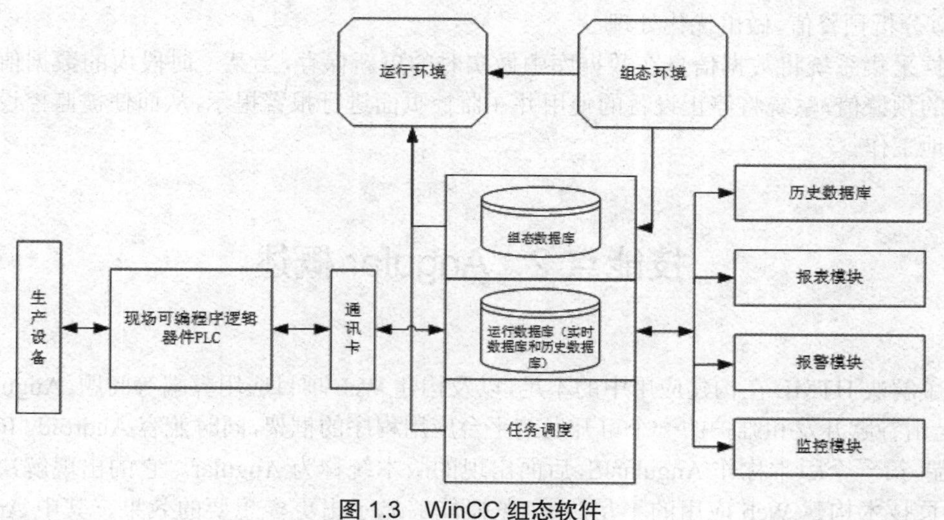

图 1.3　WinCC 组态软件

（2）数据的分析与处理

通过智慧工厂中央管理系统对采集到的数据进行处理，将其显示在页面中，并对数据进行分析，对于超过预警值的数据会进行页面报警处理，而对于普通数据则会进行分析并编写方法对其进行统计。

（3）数据的输出

数据显示在页面的过程中，使用了图表的形式，包括柱状图、列表等。对数据进行多种形式的分析，以便清晰具体地显示每组数据，让用户能够更加直观的分析出当前各个设备的使用状态及生产状态。

3　智慧工厂中央管理系统的优势

(1)图表形式展现数据

本系统使用了柱状图和列表等图表形式对数据进行描述,其中柱状图可以比较两个或两个以上的分组数据,利用图表可以更直观的反映问题。如图 1.4 所示。

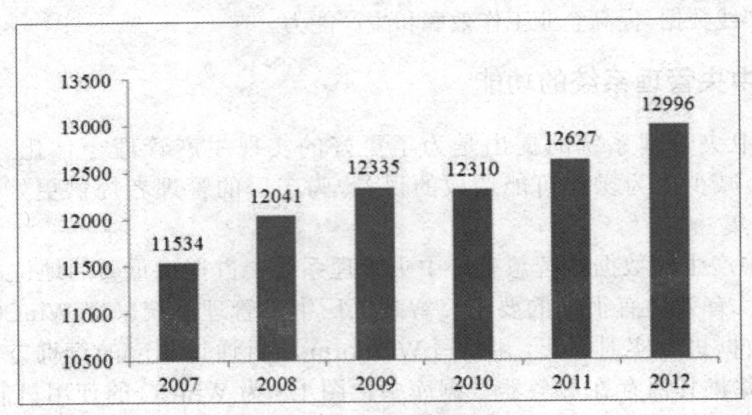

图 1.4　项目效果图

(2)分析预警值,做出优化处理

数据采集系统将数据信息在数据库中做实时的更新保存,当某一时段内的数据值超过系统认定的预警值,系统将停止设备的使用并在监控页面进行报警提示,从而使被监控设备能够高效率地工作。

技能点 2　Angular 概述

为了解决 HTML 在构建应用中的不足,以及构建 Web 项目应用弊端等问题,Angular 问世了,它是由谷歌开发和维护的一个可开发跨平台应用程序的框架,同时兼容 Android、IOS 系统和 PC 端,第一个版本称作 AngularJS,后面出现的版本统称为 Angular。它的出现解决了使用静态网页技术构建 Web 应用的不足,让浏览器能够显示出更多想要的效果。其中 AngularJS 具有的特点如下:

- 数据的双向绑定:数据的双向绑定是指 View 层的数据和 Model 层的数据是双向绑定的,其中一个发生变化另一个也会随着变化。
- 代码模块化:每个模块的代码独立拥有自己的作用域(Scope)、模型(Model)、控制器(Controller)等。
- 强大的指令(Directive):指令作为新属性可以扩展 HTML。可以通过内置的指令来为应用添加功能,也可以自定义指令。美化了 HTML 的结构,增强了可阅读性。
- 依赖注入:可以提高代码的重用性和灵活性。
- 测试驱动开发:使用 AngularJS 开发的应用在进行单元测试和端对端测试时会比其他

开发方式开发的应用更容易，弥补了传统 JavaScript 代码难以测试和维护的缺陷。

在大家熟悉并熟练使用 AngularJS 时，Angular 问世啦，Angular 和 AngularJS 的区别如下：

- 解决了 AngularJS 的架构限制问题并提升相关的性能。
- 在快速变化的 Web 时代，Angular 支持更多的组件，并添加一些针对于移动应用的新特性。
- Angular 比之前的版本开发接口更简单。
- 不再有 Controller 和 Scope。
- 引入了 RxJS 与 Observable。
- 引入了 Zone.js，提供更加智能的变化检测。

技能点 3　Angular 环境搭建

1　Node.js 安装

Node.js 是基于 Chrome JavaScript 运行时建立的平台，用于搭建易于扩展的网络应用。Node.js 对一些特殊用例进行优化，提供相应的 API。下面介绍 Node.js 的安装过程：

（1）下载安装包

在 Node.js 官网下载安装包，官网地址为：https://nodejs.org/en/，如图 1.5 所示。

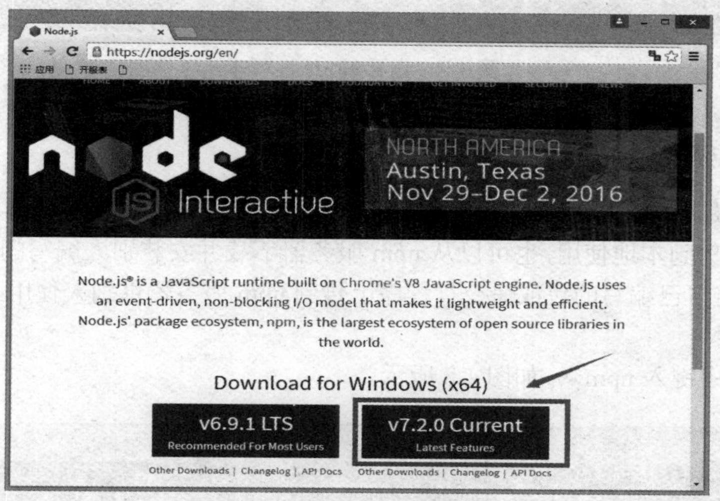

图 1.5　Node.js 官网

（2）安装 Node.js

双击打开 nodejs.exe 安装文件进行安装，如图 1.6 所示。

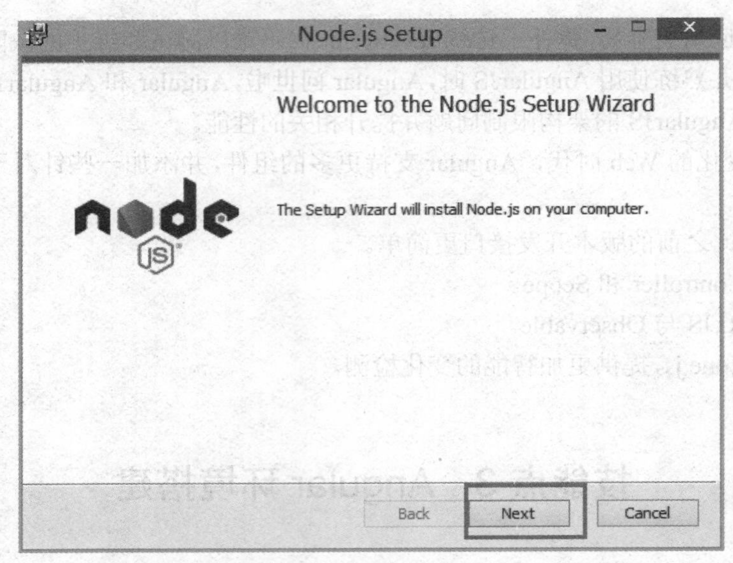

图 1.6 Node.js 安装

（3）检测 Node.js 是否安装成功

打开命令窗口，输入 node -v 显示当前版本号，表示安装成功。如图 1.7 所示。

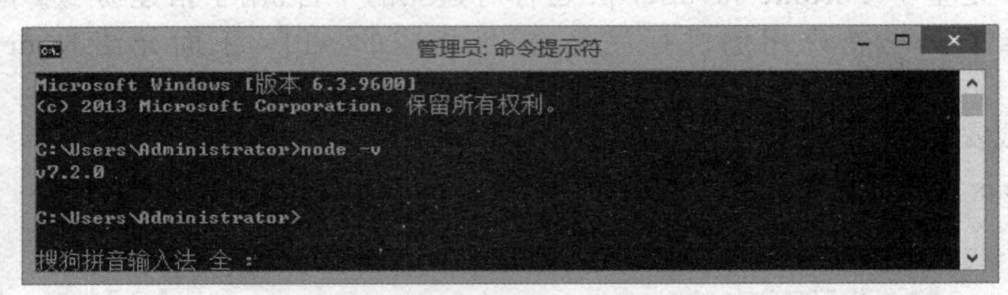

图 1.7 查看 Node.js 版本

2 Node Package Manager

npm 的全称是 Node Package Manager，是随同 Node.js 安装的包管理工具。可以从 npm 服务器下载第三方包到本地使用，也可以从 npm 服务器下载并安装别人编写的命令行程序到本地使用，还可以将自己编写的包或命令行程序上传到 npm 服务器供别人使用。

（1）查看版本

在命令窗口中键入 npm -v，如图 1.8 所示。

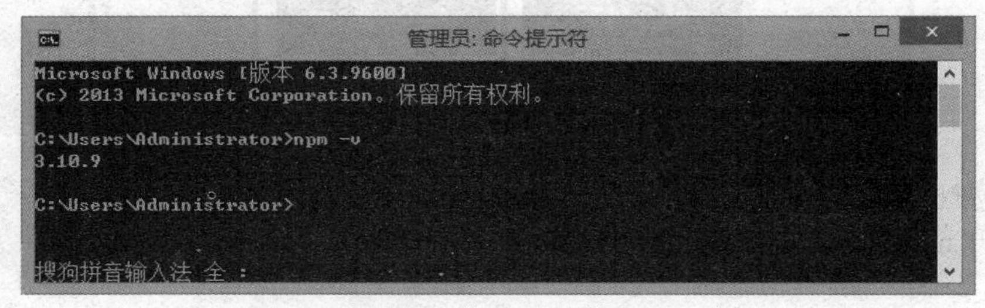

图 1.8 查看 npm 版本

（2）新建文件夹

在 nodejs 下建立 node_global 和 node_cache 两个文件夹。

（3）改变 npm 的启动和缓存位置

通过 cmd 打开命令窗口输入 npm config set prefix 和 npm config set cache 两个命令，通过这两条命令改变 npm 的启动和缓存位置，如图 1.9 和图 1.10 所示。

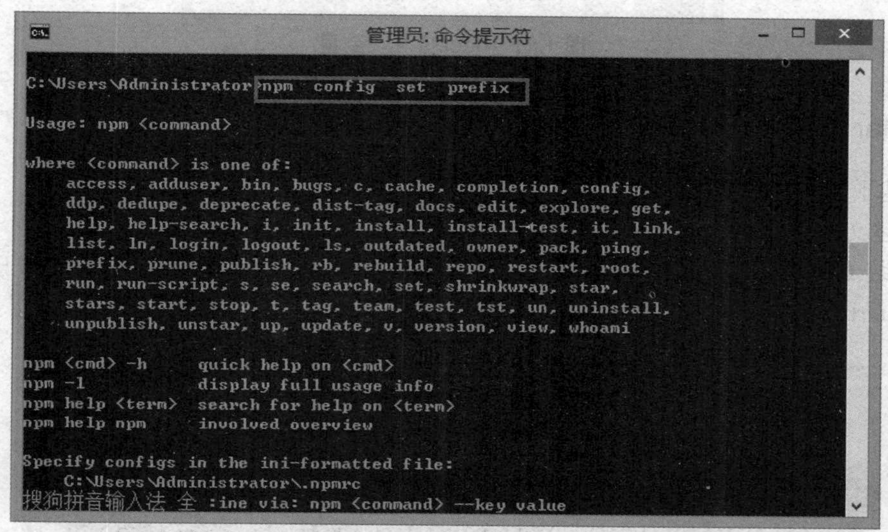

图 1.9　npm config set prefix

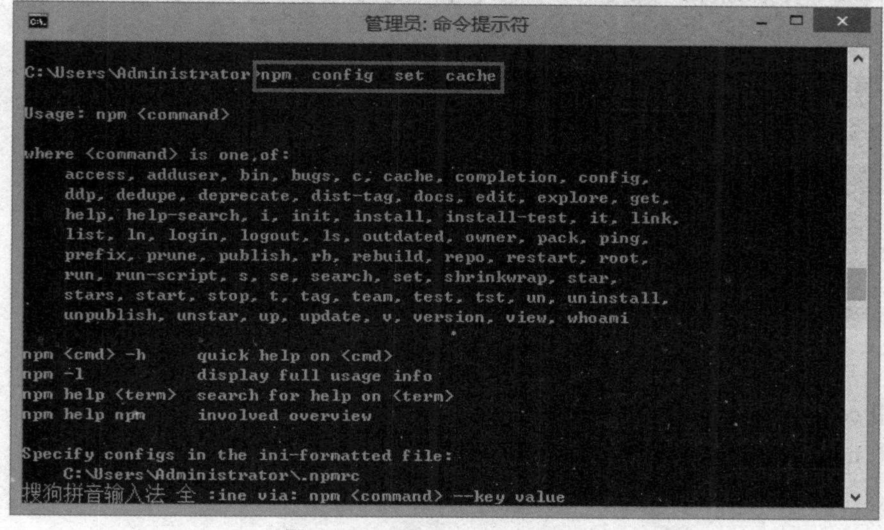

图 1.10　npm config set cache

（4）配置环境变量

新建系统变量，变量名为：NODE_PATH；变量值为：nodejs 文件安装目录，如图 1.11 所示。

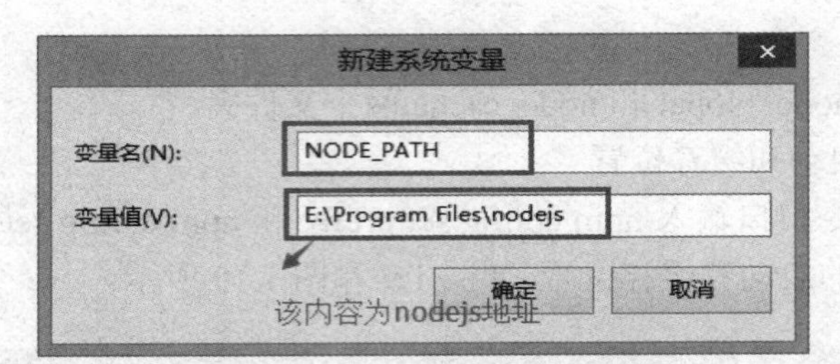

图 1.11　nodejs 环境变量

3　Angular 安装

（1）基于 Angular Quickstart

①新建文件夹，命名 project。

②安装 Git 软件（安装步骤省略）。使用 Git 克隆 quickstart 项目。文件夹中会生成 ng4-quickstart 文件夹。如图 1.12 所示。

图 1.12　克隆项目

③使用 cd ng4-quickstart 命令进入项目文件夹。

④使用 npm install 命令安装项目所需依赖。效果如图 1.13 所示。

```
管理员: C:\Windows\system32\cmd.exe
|   | `-- uuid@3.1.0
|   `-- semver@5.4.1
+-- rimraf@2.6.1
+-- rxjs@5.0.1
| `-- symbol-observable@1.0.4
+-- systemjs@0.19.40
| `-- when@3.7.8
+-- tslint@3.15.1
| +-- diff@2.2.3
| +-- findup-sync@0.3.0
| | `-- glob@5.0.15
| `-- underscore.string@3.3.4
|   +-- sprintf-js@1.1.1
|   `-- util-deprecate@1.0.2
+-- typescript@2.1.6
`-- zone.js@0.8.17

npm WARN          SKIPPING OPTIONAL DEPENDENCY: fsevents@^1.0.0 (node_modules\ch
okidar\node_modules\fsevents):
npm WARN          SKIPPING OPTIONAL DEPENDENCY: Unsupported platform for fsevents@
1.1.2: wanted {"os":"darwin","arch":"any"} (current: {"os":"win32","arch":"x64"}
)

C:\Users\Administrator.USER-20170510ZB\Desktop\project\ng4-quickstart>
        半:
```

图 1.13　安装依赖

⑤运行 npm start 来启动该项目，效果如图 1.14 所示。并通过浏览器打开一个窗口，效果如图 1.15 所示，则说明 Angular 安装成功。

```
lite-server
[1] 17.09.06 16:16:41 304 GET /rxjs/Subscriber.js
[1] 17.09.06 16:16:41 304 GET /rxjs/Subscription.js
[1] 17.09.06 16:16:41 304 GET /rxjs/util/ObjectUnsubscribedError.js
[1] 17.09.06 16:16:41 304 GET /rxjs/SubjectSubscription.js
[1] 17.09.06 16:16:41 304 GET /rxjs/symbol/rxSubscriber.js
[1] 17.09.06 16:16:41 304 GET /rxjs/Observer.js
[1] 17.09.06 16:16:41 304 GET /rxjs/observable/ArrayObservable.js
[1] 17.09.06 16:16:41 304 GET /rxjs/operator/mergeAll.js
[1] 17.09.06 16:16:41 304 GET /rxjs/util/isScheduler.js
[1] 17.09.06 16:16:41 304 GET /rxjs/observable/ConnectableObservable.js
[1] 17.09.06 16:16:41 304 GET /rxjs/util/isFunction.js
[1] 17.09.06 16:16:41 304 GET /rxjs/util/isArray.js
[1] 17.09.06 16:16:41 304 GET /rxjs/util/isObject.js
[1] 17.09.06 16:16:41 304 GET /rxjs/util/tryCatch.js
[1] 17.09.06 16:16:41 304 GET /rxjs/util/errorObject.js
[1] 17.09.06 16:16:41 304 GET /rxjs/util/UnsubscriptionError.js
[1] 17.09.06 16:16:41 304 GET /rxjs/observable/ScalarObservable.js
[1] 17.09.06 16:16:41 304 GET /rxjs/observable/EmptyObservable.js
[1] 17.09.06 16:16:41 304 GET /rxjs/OuterSubscriber.js
[1] 17.09.06 16:16:41 304 GET /rxjs/util/subscribeToResult.js
[1] 17.09.06 16:16:41 304 GET /rxjs/util/isPromise.js
[1] 17.09.06 16:16:41 304 GET /rxjs/symbol/iterator.js
[1] 17.09.06 16:16:41 304 GET /rxjs/InnerSubscriber.js
_
        半:
```

图 1.14　启动项目界面

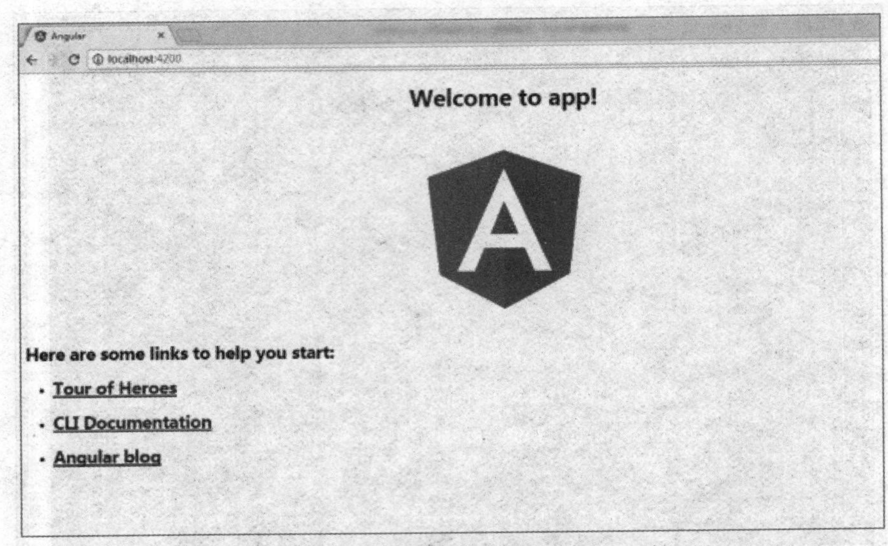

图 1.15 运行成功界面

（2）基于 Angular CLI

①打开 project 文件夹，按住“shift”+“鼠标右键”，出现弹出框，选择“在此处打开命令窗口”，弹出如图 1.16 所示界面。

图 1.16 命令窗口界面

②使用 npm install -g @angular/cli 命令安装 Angular CLI。效果如图 1.17 所示。

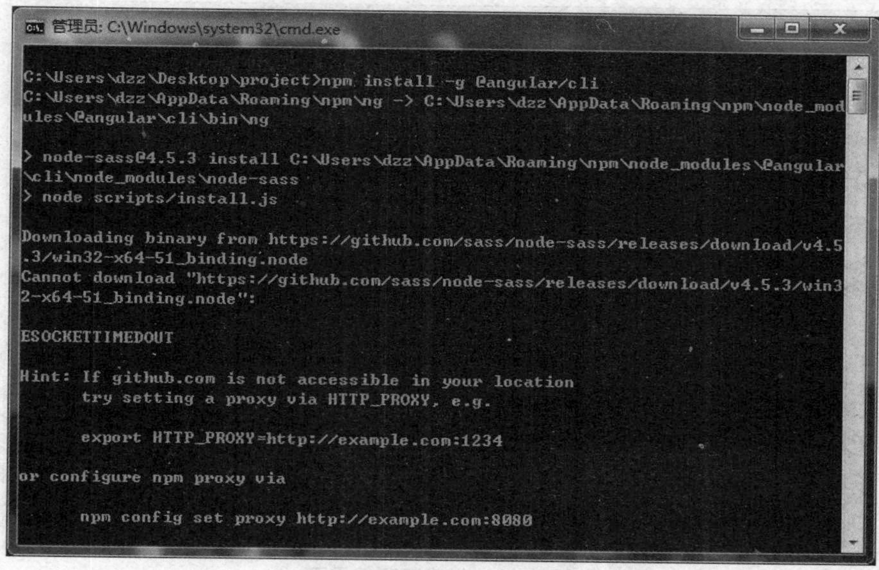

图 1.17　安装 Angular CLI

③使用 ng --version 检测是否安装成功，出现如图 1.18 所示效果，则说明安装成功。

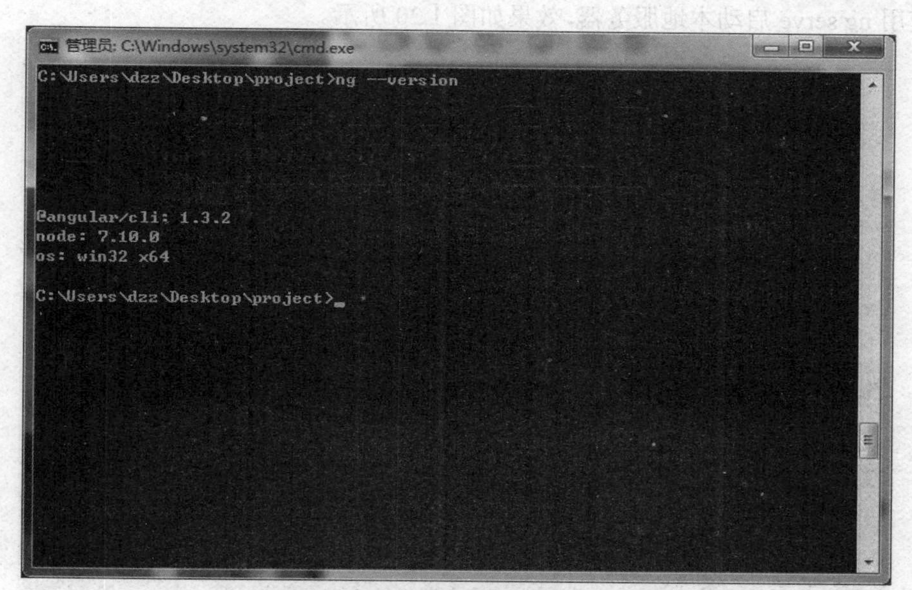

图 1.18　检测是否安装成功

④使用 ng new projectname 命令创建新项目，效果如图 1.19 所示。

图 1.19 创建新项目界面

⑤使用 cd projectname 命令进入 projectname 目录。

⑥使用 ng serve 启动本地服务器，效果如图 1.20 所示。

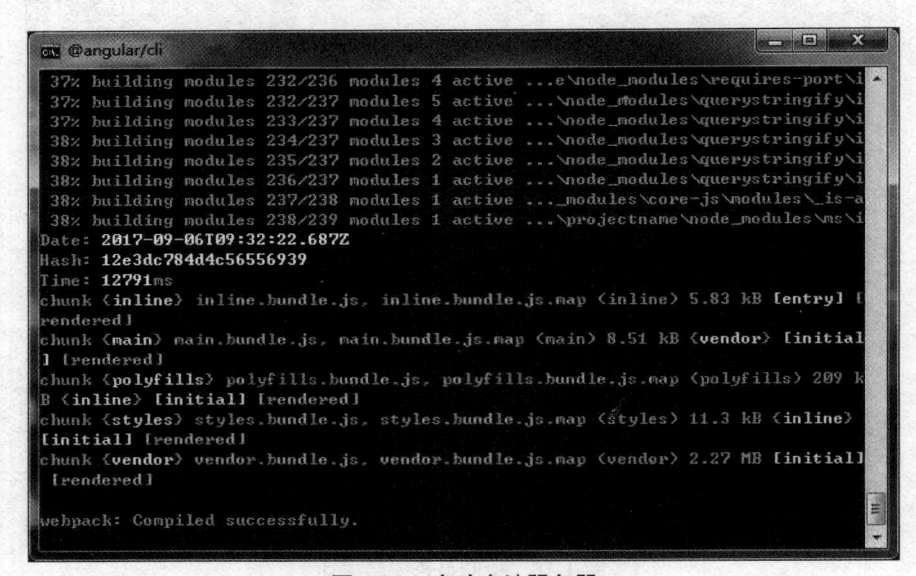

图 1.20 启动本地服务器

⑦在浏览器中输入 http://localhost:4200，出现如图 1.21 所示效果，则说明 Angular 环境搭建成功。

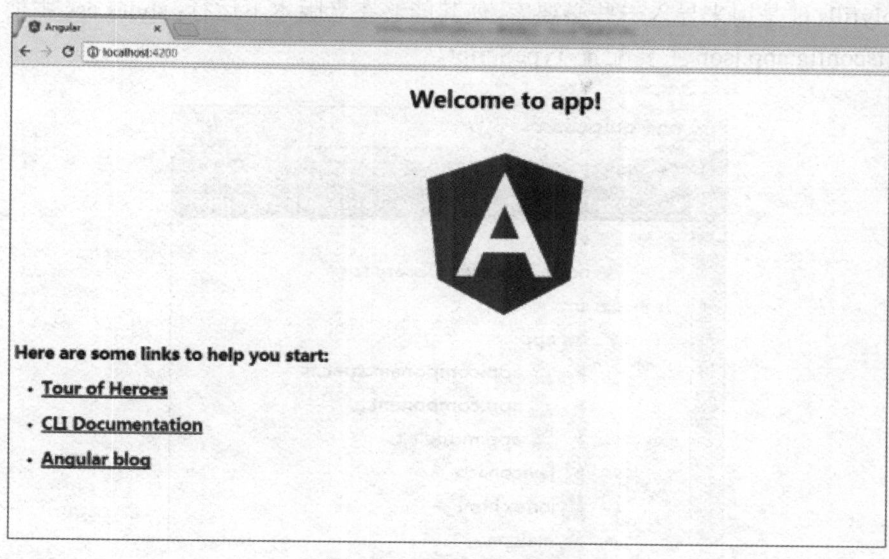

图 1.21　运行成功效果

快来扫一扫！

提示：当对 Angular 所需环境有一定了解后，你是否打算放弃本门课程的学习呢？扫描图中二维码，你的想法是否有所改变呢？

技能点 4　Angular 项目结构

当 Angular 环境安装完成后，可以看到两种方式安装后的结构目录基本一致（此处讲解基于 Angular Quickstart 项目结构），那么 Angular 所对应的目录都有什么作用呢？Angular 目录结构如图 1.22 所示。

通过目录结构图可以看出 Angular 文件分为几层，其中最外层包括 e2e、node_modules、src 以及一些配置文件，这些文件的主要功能如表 1.1 所示。

除了最外层文件，在文件夹 src 目录下也存在一些文件夹和文件，src 文件夹是项目的一个核心，其中 app 目录包含应用的组件和模块，在开发项目中所有的编码都在该文件下；assets 是存储静态资源，比如视频图片等；environments 主要用于环境的配置；index.html 为整个应用的根 HTML，程序启动时访问该界面；main.ts 为整个项目的入口点，Angular 通过这个文件来启动项目；polyfills.ts 是用来导入一些必要库，使其能够在老版本下运行；styles.css 主要是放置全局的样式；tsconfig.app.json 主要配置 TypeScript。

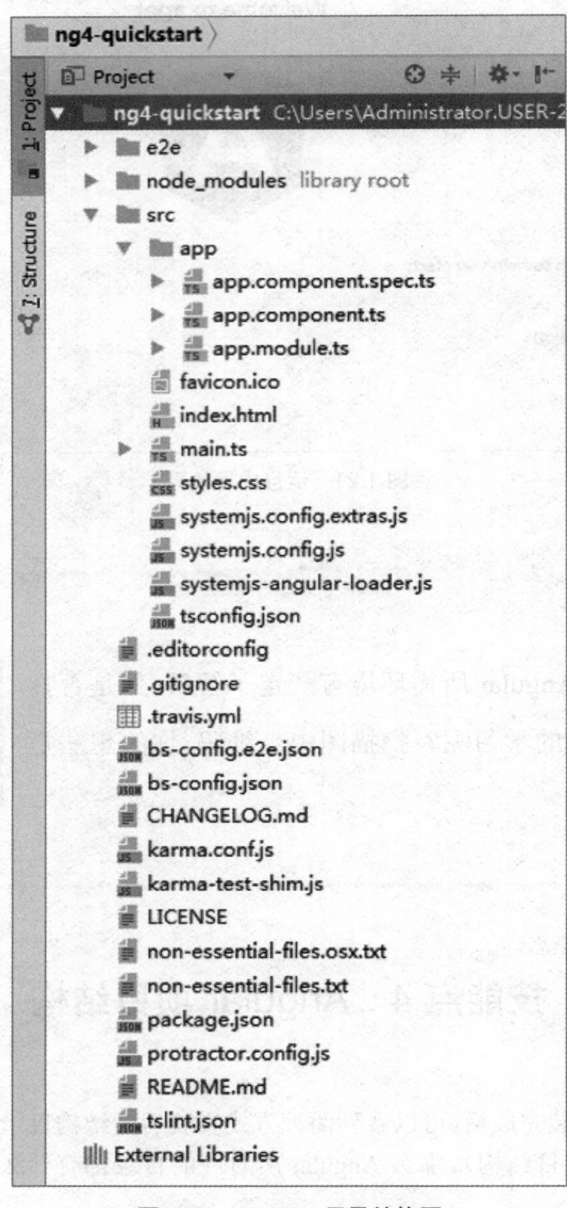

图 1.22　Angular 目录结构图

表 1.1　Angular 最外层文件夹作用

文件名称	描述
e2e	端到端的测试目录　用来做自动测试
node_modules	第三方依赖包存放目录，通常使用 npm install 安装的依赖都存在该文件夹下
src	存放项目文件
.Angular-cli.json	Angular 命令行工具的配置文件。可以引用其他的第三方包，如 jQuery 等
karma.conf.js	karma 的配置文件
package.json	该文件里面列出了该应用程序所使用的第三方依赖包
protractor.conf.js	做自动化测试的配置文件
README.md	说明文件
tslint.json	tslint 的配置文件，用来定义 TypeScript 代码质量检查的规则

整体项目结构如图 1.23 所示。

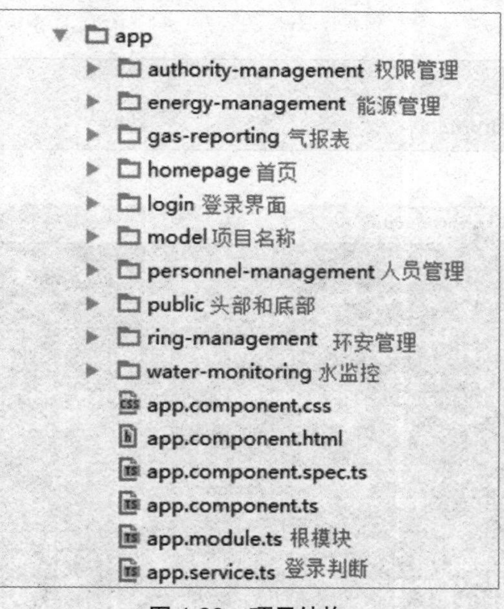

图 1.23　项目结构

通过下面十三个步骤的操作，实现图 1.2 所示的智慧工厂中央管理系统登录模块的效果。

第一步：创建 project 项目。通过在命令窗口输入命令创建智慧工厂中央管理系统项目，命令如下所示。创建成功界面如图 1.24 所示。

```
ng new project
```

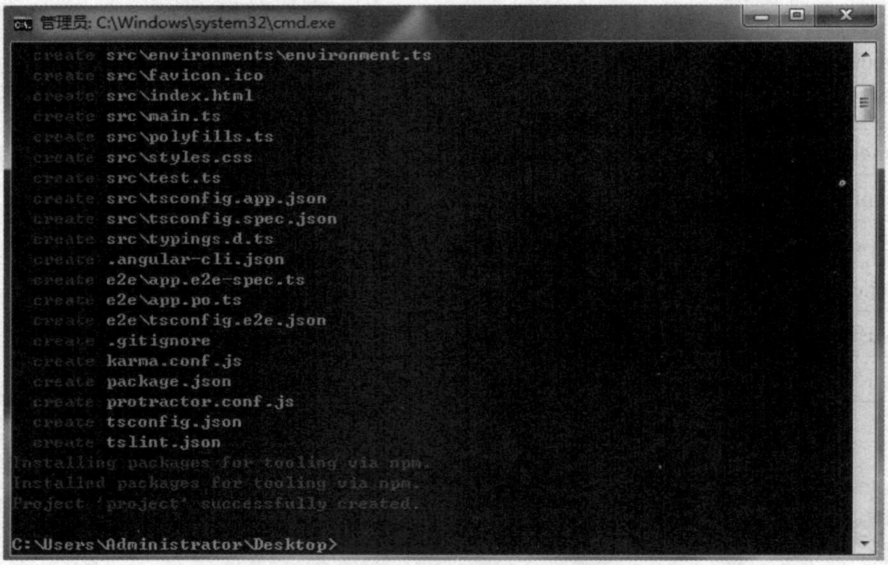

```
管理员: C:\Windows\system32\cmd.exe
  create src\environments\environment.ts
  create src\favicon.ico
  create src\index.html
  create src\main.ts
  create src\polyfills.ts
  create src\styles.css
  create src\test.ts
  create src\tsconfig.app.json
  create src\tsconfig.spec.json
  create src\typings.d.ts
  create .angular-cli.json
  create e2e\app.e2e-spec.ts
  create e2e\app.po.ts
  create e2e\tsconfig.e2e.json
  create .gitignore
  create karma.conf.js
  create package.json
  create protractor.conf.js
  create tsconfig.json
  create tslint.json
Installing packages for tooling via npm.
Installed packages for tooling via npm.
Project 'project' successfully created.

C:\Users\Administrator\Desktop>
```

图 1.24 创建项目

第二步：安装第三方依赖。在命令窗口输入如下命令，安装 jQuery 和 Bootstrap 库，效果图如图 1.25 所示。

```
npm install jquery --save
npm install bootstrap --save
```

```
管理员: C:\Windows\system32\cmd.exe
C:\Users\Administrator\Desktop\project>npm install jquery --save
project@0.0.0 C:\Users\Administrator\Desktop\project
`-- jquery@3.2.1

npm WARN optional SKIPPING OPTIONAL DEPENDENCY: fsevents@^1.0.0 (node_modules\ch
okidar\node_modules\fsevents):
npm WARN notsup SKIPPING OPTIONAL DEPENDENCY: Unsupported platform for fsevents@
1.1.2: wanted {"os":"darwin","arch":"any"} (current: {"os":"win32","arch":"x64"}
)

C:\Users\Administrator\Desktop\project>npm install bootstrap --save
project@0.0.0 C:\Users\Administrator\Desktop\project
`-- bootstrap@3.3.7

npm WARN optional SKIPPING OPTIONAL DEPENDENCY: fsevents@^1.0.0 (node_modules\ch
okidar\node_modules\fsevents):
npm WARN notsup SKIPPING OPTIONAL DEPENDENCY: Unsupported platform for fsevents@
1.1.2: wanted {"os":"darwin","arch":"any"} (current: {"os":"win32","arch":"x64"}
)

C:\Users\Administrator\Desktop\project>
```

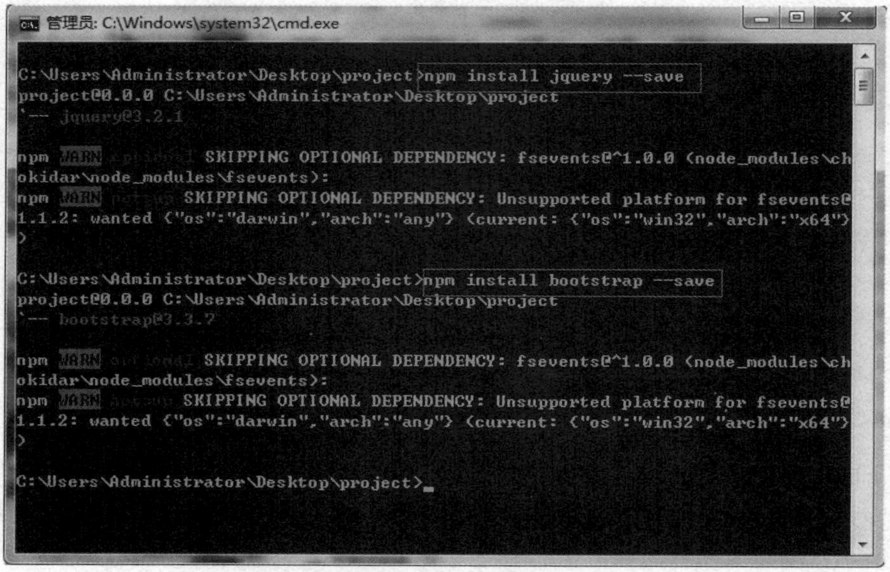

图 1.25 安装依赖

第三步：打开 project 项目，在 .Angular-cli.json 文件中配置 jQuery 和 Bootstrap 路径，代码如下所示。

```
"styles": [
        "styles.css",
        "../node_modules/bootstrap/dist/css/bootstrap.css",
            ],
"scripts": [
        "../node_modules/jquery/dist/jquery.js",
        "../node_modules/bootstrap/dist/js/bootstrap.js"
            ],
```

第四步：安装 Bootstrap 和 jQuery 类型描述文件，在命令窗口输入如下命令，效果如图 1.26 所示。

```
npm install @types/bootstrap --save-dev
npm install @types/jquery --save-dev
```

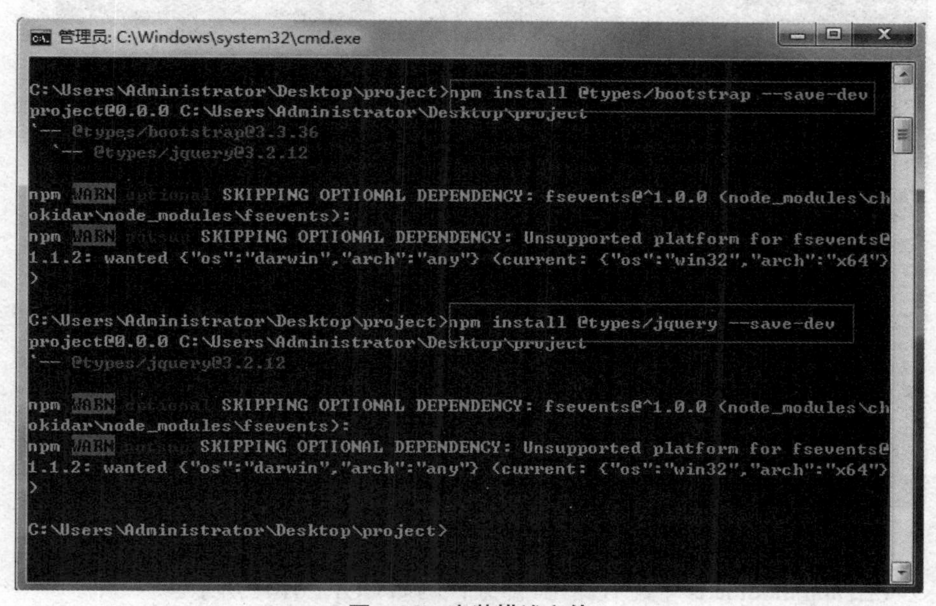

图 1.26　安装描述文件

第五步：将登录界面分成左边和右边两部分，通过命令创建主组件以及左侧标题和右侧表单输入组件，在根目录下创建服务，命令如下所示。效果如图 1.27 所示。

```
ng g component login
ng g component loginleft
ng g component loginright
ng g service app
```

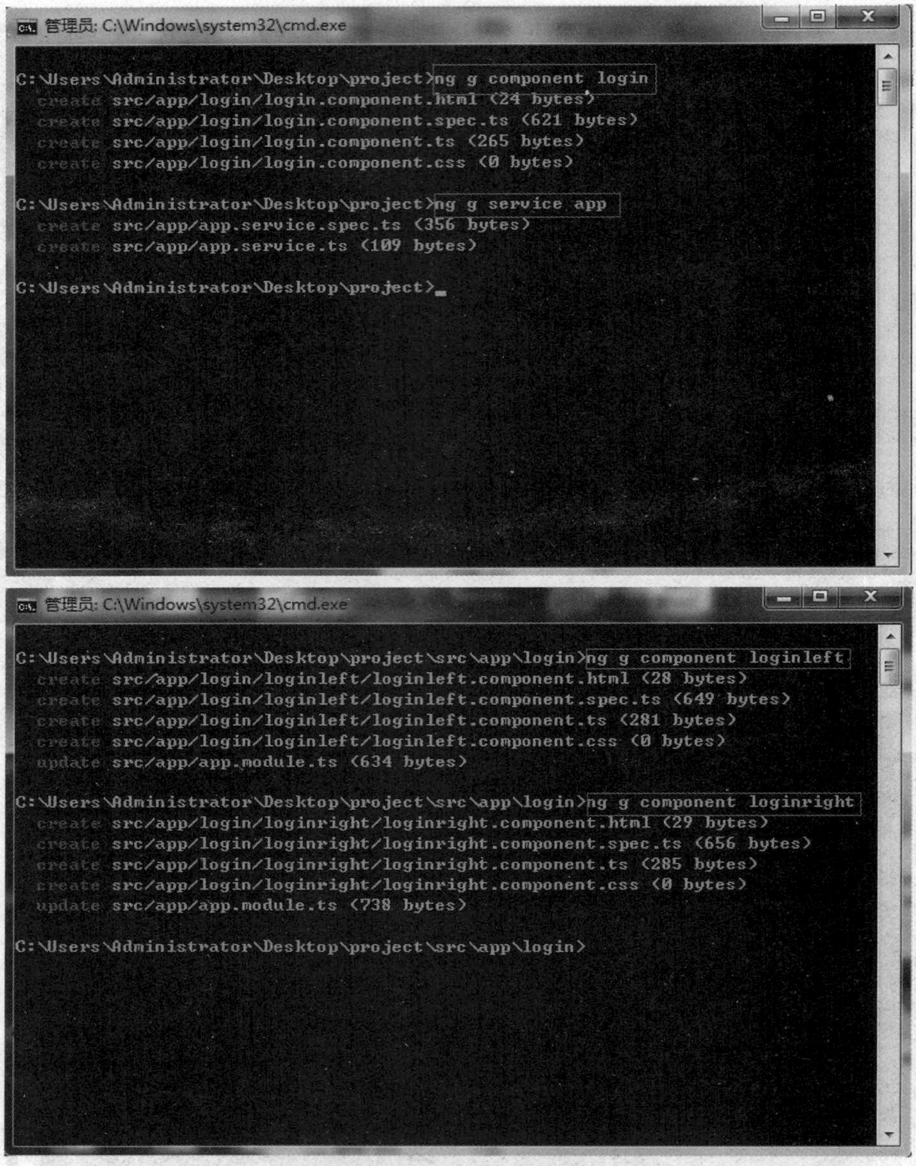

```
管理员: C:\Windows\system32\cmd.exe

C:\Users\Administrator\Desktop\project>ng g component login
  create src/app/login/login.component.html (24 bytes)
  create src/app/login/login.component.spec.ts (621 bytes)
  create src/app/login/login.component.ts (265 bytes)
  create src/app/login/login.component.css (0 bytes)

C:\Users\Administrator\Desktop\project>ng g service app
  create src/app/app.service.spec.ts (356 bytes)
  create src/app/app.service.ts (109 bytes)

C:\Users\Administrator\Desktop\project>
```

```
管理员: C:\Windows\system32\cmd.exe

C:\Users\Administrator\Desktop\project\src\app\login>ng g component loginleft
  create src/app/login/loginleft/loginleft.component.html (28 bytes)
  create src/app/login/loginleft/loginleft.component.spec.ts (649 bytes)
  create src/app/login/loginleft/loginleft.component.ts (281 bytes)
  create src/app/login/loginleft/loginleft.component.css (0 bytes)
  update src/app/app.module.ts (634 bytes)

C:\Users\Administrator\Desktop\project\src\app\login>ng g component loginright
  create src/app/login/loginright/loginright.component.html (29 bytes)
  create src/app/login/loginright/loginright.component.spec.ts (656 bytes)
  create src/app/login/loginright/loginright.component.ts (285 bytes)
  create src/app/login/loginright/loginright.component.css (0 bytes)
  update src/app/app.module.ts (738 bytes)

C:\Users\Administrator\Desktop\project\src\app\login>
```

图 1.27　新建组件与服务

第六步：在 app.module.ts 中配置路由，部分代码如 CORE0101 所示。

代码 ORE0101：配置路由

```
import { RouterModule，Routes } from '@angular/router';
import { HttpModule } from '@angular/http';
// 配置路由
const appRoutes: Routes = [
  { path: 'login', component: LoginComponent },
```

```
];
imports: [
  HttpModule,
  RouterModule.forRoot(appRoutes),
],
```

第七步：在 app.component.ts 文件中设置 router-outlet，定义路由显示位置。代码如下所示。

```
<router-outlet></router-outlet>
```

第八步：在 login.component.html 文件中进行界面布局，并设置左侧标题组件以及右侧表单输入组件渲染位置，代码如 CORE0102 所示。

代码 CORE0102：界面布局

```
<div class="row login-window fixed-top" >
 <div class="row conent" >
   <!-- 左侧页面 -->
   <app-loginleft></app-loginleft>
   <!-- 右侧输入窗口 -->
   <app-loginright></app-loginright>
  </div>
</div>
```

在 login.component.css 文件中为其设置背景图片等样式，代码如 CORE0103 所示，效果如图 1.28 所示。

代码 CORE0103：设置样式

```
.login-window {
  height: 100%;
  background-size: 100% 100%;
  background-image: url(../../assets/welcome.jpg) ;
  padding-top: -10px;
}
.fixed-top {
  position: fixed;
  right: 0;
  left: 0;
  z-index: 1030;
}
```

```
.login-window{
   margin-top: -110px
}
.conent{
   padding-left: 250px;
   padding-top: 150px
}
```

图 1.28　界面布局

第九步：设置左侧标题内容，通过 h2 标签设置智慧工厂中央管理系统的标题，代码如 CORE0104 所示，设置样式前效果如图 1.29 所示。

代码 CORE0104：左侧标题

```
<div class="bg col-lg-5" >
   <h2 > 智慧工厂中央管理系统 </h2>
</div>
```

图 1.29　设置样式前

在 loginleft.component.css 文件中为标题设置背景，并居中显示，代码如 CORE0105 所示，设置样式后效果如图 1.30 所示。

代码 CORE0105：设置样式

```
.bg{
  opacity: .85;
  background-color: #f2f2f2;
  text-align: center;
  width: 350px;
  height: 280px
}
 .bg h2{
  line-height: 200px
}
```

图 1.30　设置样式后

第十步：使用表单设置用户名、密码输入，在 form 标签中创建 doLogin() 方法，通过 [disabled] 设置登录按钮禁用效果。代码如 CORE0106 所示。设置样式前效果如图 1.31 所示。

代码 CORE0106：表单输入

```
<div class="card-block col-lg-7" >
  <div style="padding: 20px;height: 280px">
    <h4 class="card-title" >
      <img src="../../assets/login_icon.png">    请登录 !
    </h4>
    <p class="text-muted">
      <small > 请输入管理员账号密码 </small>
      <small class="text-danger pull-left">
```

```
        <i class="fa fa-warning" ></i>
      </small>
    </p>
    <!-- 登录表单 -->
    <form #loginForm="ngForm" (ngSubmit)= "doLogin()">
      <div class="input-group">
        <span class="input-group-addon rounded-0"><i class="fa fa-user"></i></span>
        <input type="text" name="login_account" #login_account="ngModel"
               [(ngModel)]="account" required class="form-control rounded-0"
               placeholder=" 账号 ">
    </div>
    <br>
    <div class="input-group">
      <span class="input-group-addon rounded-0"><i class="fa fa-lock"></i></span>
      <input type="text" name="login_password" #login_password="ngModel"
             [(ngModel)]="password" required class="form-control rounded-0"
             placeholder=" 密码 ">
    </div>
    <br>
    <p class="card-text">
      <button class="btn btn-info" type="submit" [disabled]="!loginForm.form.valid">
        <i class="fa fa-sign-in"></i> 登录 </button>
      <span class="pull-right"> 忘记密码？ </span>
    </p>
   </form>
  </div>
</div>
```

图 1.31　设置样式前

在 loginright.component.css 文件中为其设置黑色的背景，并设置字体颜色、大小及显示位置等。代码如 CORE0107 所示，设置样式后效果如图 1.32 所示。

代码 CORE0107：设置样式

```
.card-block{
  background-color: black;
  width: 450px
}
.card-block .card-title{
  color: #ffffff
}
.card-block .text-muted small{
  color:red
}
.card-text{
  float: left;
}
.pull-right{
  color: #ffffff;
  padding-left: 200px;
}
```

图 1.32　设置样式后

第十一步：在 app.service.ts 文件中获取用户名与密码输入值。代码如 CORE0108 所示。

代码 CORE0108：登录校验

```
import { Injectable } from '@angular/core';
```

```
import { Router } from '@angular/router';
@Injectable()
export class AppService {
    constructor(public router: Router) {}
    login(account: string, password: string): void {
      }
}
```

第十二步：在 login.component.ts 文件中使用依赖注入，注入服务，通过 doLogin() 方法获取数据，并传递给表单，当用户名和密码同时存在时，登录按钮禁用效果消除，代码如 CORE0109 所示。

代码 CORE0109：调用服务

```
import { Component, OnInit } from '@angular/core';
// 依赖注入 AppService 服务
import {AppService} from "../app.service";
@Component({
  selector: 'app-login',
  templateUrl: './login.component.html',
  styleUrls: ['./login.component.css'],
  providers: [AppService]
})
export class LoginComponent implements OnInit {
  // 声明数据类型
  public account: string;
  public password: string;
  // 构造器
  constructor(public appservice: AppService) {
  }
  login(account:string, password:string): void {
    this.appservice.login(account, password);}
  ngOnInit() {}
  doLogin(): void {
    this.appservice.login(this.account, this.password);
  }
}
```

第十三步：在 app.service.ts 文件中对每次用户输入信息进行校验，当用户名与密码输入值不为空时，点击登录按钮跳转到主页面。代码如 CORE0110 所示。

代码 CORE0110：登录校验

```
import { Injectable } from '@angular/core';
import { Router } from '@angular/router';
@Injectable()
export class AppService {
  constructor(public router: Router) {}
  // 每次用户登录进行校验
  login(account: string, password: string): void {
    if(account && password) {
      this.router.navigateByUrl('/ 主页面（自己设置 '));
    }
    else {
      this.router.navigateByUrl(‘/login’);
    }
  }
}
```

至此，智慧工厂中央管理系统登录模块制作完成。

本项目通过对智慧工厂中央管理系统登录模块的学习，对 Bootstrap 和 Angular 的作用及发展具有初步了解，能够使用不同方式搭建 Angular 环境及掌握 Angular 项目结构，同时也加深了使用 Bootstrap 对界面进行布局和美化的印象。

Bootstrap	前端库
Observable	可观察量
scope	作用域
Model	模型
Directive	指令
Node Package Manager	包管理工具
Node Version Manager Nodejs	版本管理工具
global	全球的
version	版本

一、选择题

1. Angular 是由（ ）公司推出的。

（A）Twitter （B）Titanic （C）Tenga （D）Google

2.（ ）是 Angular 的最初版本。

（A）AngularJS （B）2.x （C）Bootstrap 4 alpha （D）4.x

3. 下面对 Angular 说法错误的是（ ）。

（A）具有 Controller 和 Scope

（B）引入了 RxJS 与 Observable

（C）引入了 Zone.js，提供更加智能的变化检测

（D）Angular 比之前的版本开发接口更简单

4. 下面描述正确的是（ ）。

（A）e2e：配置文件

（B）karma.conf.js：Angular 的配置文件

（C）tslint.json：测试文件

（D）protractor.conf.js：做自动化测试的配置文件

5. 下面描述错误的是（ ）。

（A）assets：是存储静态资源，比如视频图片等

（B）styles.css：设置组件样式

（C）index.html：程序启动时访问该界面

（D）environments：主要用于环境的配置

二、填空题

1. 安装项目所需依赖的命令是 _________。

2. 创建新项目的命令是 _________。

3. 启动项目的命令是 _________。

4. Angular 项目中的 node_modules 文件起什么作用 _________。

5. 在 Angular-cli.json 文件中可以 _________。

三、上机题

通过本项目所学的技能，在自己电脑上通过两种方式安装 Angular。

项目二　智慧工厂主界面

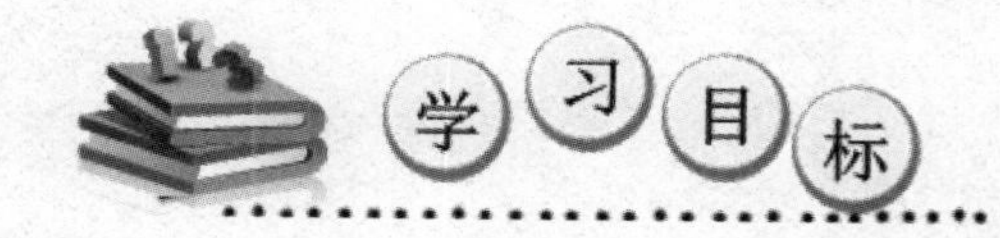

通过智慧工厂主界面模块的实现，了解主界面的基本布局及功能实现，学习 TypeScript 概述、安装，掌握如何使用 TypeScript 内置类型，具有在 TypeScript 中使用函数和类的能力。在任务实现过程中：

- 了解智慧工厂项目的基本布局。
- 学习 TypeScript 概述、安装。
- 掌握 TypeScript 内置类型。
- 具备使用函数和类的能力。

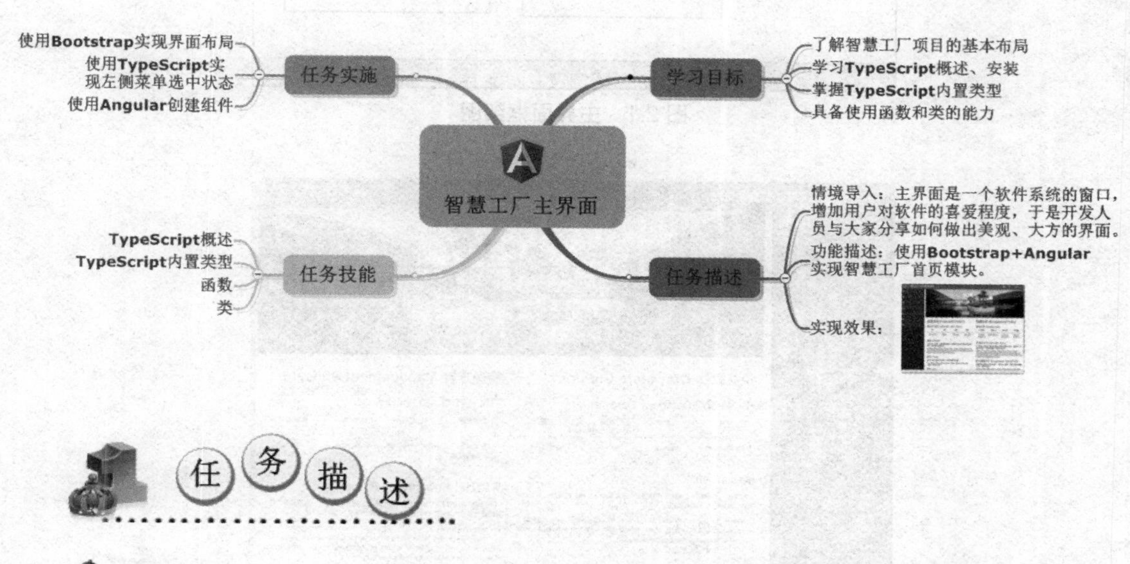

任务描述

【情境导入】

智慧工厂中央管理系统的主界面是一个软件系统的窗口，简洁美观的界面可以给用户带来一种视觉享受，增加用户对软件的喜爱程度，于是开发人员与大家分享如何做出美观、大方的界面。通过该界面，详细的介绍了公司的企业文化与管理方针，优美的布局给人眼前一亮。而且使用了 TypeScript 语言，在一定程度上提高了项目的性能。本项目主要是通过实现智慧工厂主界面来学习 TypeScript 相关知识。

【功能描述】

使用 Bootstrap+Angular 实现智慧工厂主界面：

- 使用 Bootstrap 实现界面布局。
- 使用 TypeScript 实现左侧菜单选中状态。
- 使用 Angular 创建组件。

【基本框架】

基本框架如图 2.1 所示，通过本项目的学习，能将图 2.1 的框架图转换成智慧工厂主界面，效果图如图 2.2 所示。

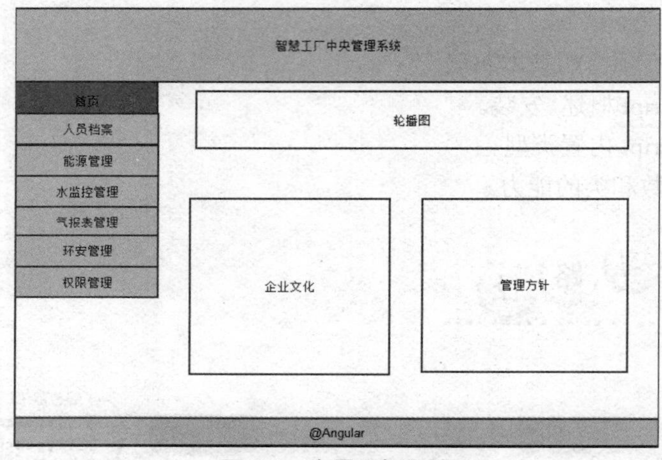

图 2.1　主界面框架图

图 2.2　主界面效果图

技能点 1　TypeScript 概述

1　TypeScript 简介

JavaScript（简称 JS）是一个脚本语言，具有使页面活灵活现等优点。由于移动端、PC 端应用的不断发展，越来越多的编程人员使用纯 JS 开发。但在编写项目时，JS 开发具有整个界面的构造过于复杂且不同版本的浏览器支持程度不同等问题。目前，仅仅依靠 JS 本身的语言特性是不能解决的，为此，微软开发了一种自由和开源的编程语言——TypeScript。

TypeScript 是 JavaScript 的超集（扩展了 JavaScript 语句），其本质是向 JavaScript 添加了静态类型（声明数据类型）和基于类的面向对象编程（如属性、方法、继承等）。具有协同开发、提高效率等优点。TypeScript 主要特点包括：

- 增加了可选类型、类和模块。
- 可编译成可读的、标准的 JavaScript。
- 支持开发大规模 JavaScript 应用。
- 用于开发大型应用，并保证编译后的 JavaScript 代码兼容性。
- 文件扩展名为 .ts，通过编辑器可编译成 .js 文件。
- 可以更加方便的调试项目。

2　TypeScript 安装

使用 TypeScript 语言之前，需要通过 Node.js 包管理器进行安装，安装步骤如下所示：

第一步：安装 Node.js（详见项目一）。

第二步：通过 Node.js 的 npm 安装 TypeScript。

在命令窗口输入如下所示命令，安装完成后效果如图 2.3 所示。

```
npm install -g typescript
```

注：如果已安装淘宝镜像，可使用 cnpm 进行安装。

```
cnpm install -g typescript
```

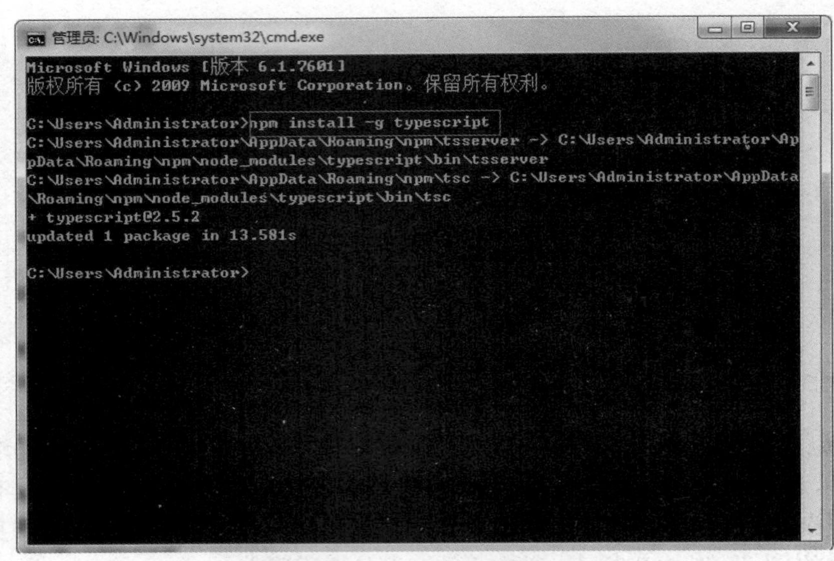

图 2.3 TypeScript 安装效果图

当安装完成后，在命令窗口输入 tsc –v 进行安装检验，如果出现图 2.4 所示，则表示安装成功。

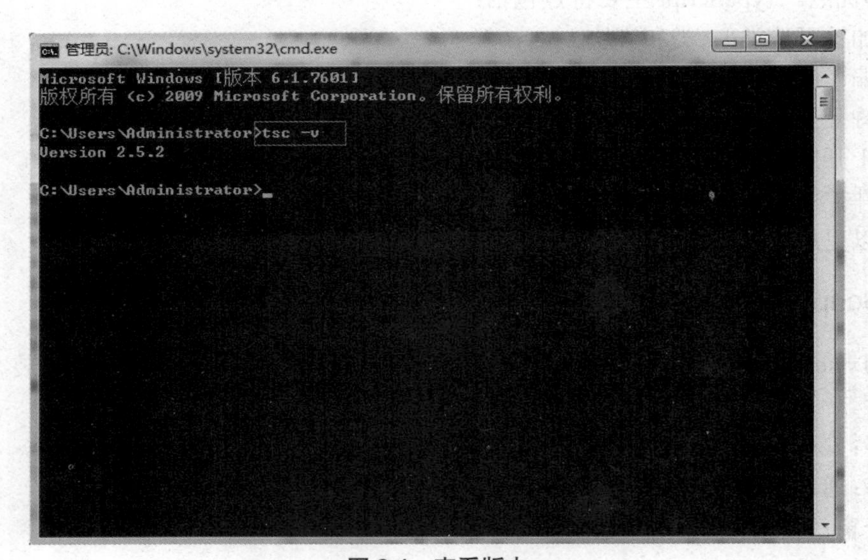

图 2.4 查看版本

提示：以下为 TypeScript 编译成 JavaScript 的案例

当安装成功后，创建一个 test.ts 文件，输入如下示例代码：

```
var t : number = 1;
```

将 test.ts 文件放入新建文件夹 test 中，在 test 文件夹中打开命令窗口，执行以下命令：

```
tsc test.ts
```

注：tsc 是 TypeScript 编译器的命令窗口接口。

通过该命令可以将 TypeScript 编译成 JavaScript 文件，最终会在 test 文件夹中生成 test.js 文件，如图 2.5 所示。

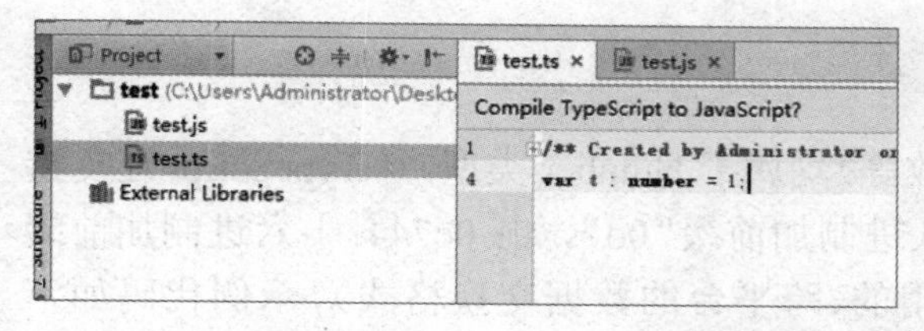

图 2.5　ts 文件转 js 文件

技能点 2　TypeScript 内置类型

在定义变量时，每个变量都必须声明类型，如 C 语言中的 Int、Boolean 类型，Java 中的 Short、Float、Char 类型等。TypeScript 也具有数据类型，其基本数据类型有 Boolean、Number、String 等。数据类型的使用具有使对象处理数据更方便、更容易被理解等优点。常用基本数据类型如表 2.1 所示。

表 2.1　基本数据类型

数据类型	关键词	描述
Boolean	boolean	布尔类型
Number	number	数字类型
String	string	字符串类型
Array	Array	数组类型
Enum	enum	枚举类型
Any	any	任意类型
Void	void	空值类型

TypeScript 支持 let（在包含它的代码块内访问）、const（一般声明常量，常量被声明后，值无法改变）、var（可在包含它的函数内或外访问）关键词声明变量，本书主要应用 let 关键词声明变量，其语法结构示例代码如下所示。

```
let 变量 : 数据类型 = 值 ;
```

1　布尔类型

目前，最简单且最常用的数据类型为布尔类型（Boolean），其对应的属性值只能为 true 或

false，示例代码如下所示。

```
let stu: boolean = false;
```

2 数字类型

在 TypeScript 中包含的数字（浮点数、整数等）均属于 Number 类型，且 TypeScript 还支持二进制等（二进制加前缀“0b”，如：0b1010；八进制加前缀“0o”，如：0o744；十六进制加前缀“0x”，如：0xf00d）字面量（字面量是一种通用的、跨平台的数据交换格式），示例代码如下所示。

```
let id: number = 9;
let Floor _nub: number = 0b1010;
```

3 字符串类型

在 C、Java、PHP 等编程语言中，字符串（String）应用很常见，TypeScript 也不例外，其使用字符串作为数据类型。实现方式是通过双引号（" "）或单引号（' '）将字符串值进行包裹，示例代码如下所示。

```
let name: string = "Michelle ";
name = 'James';
```

在使用 TypeScript 定义变量或设置相关样式及属性时，往往使用字符串嵌套，正确代码如下所示。

```
let name: string = "Hello, my name is 'Lin' ";
let name: string ='Hello, my name is "Lin" ';
```

错误代码如下所示。

```
let name: string = "Hello, my name is "Lin" ";
let name: string ='Hello, my name is 'Lin' ';
```

字符串中包含模板字符串，主要作用是定义多行文本和内嵌表达式，其使用方式是通过反引号（` `）将字符串值进行包裹，并以 ${ 表达式 } 形式嵌入表达式，示例代码如下所示。

```
let age: number = 7;
let say: string = `I’ m ${ age} years old.`;
console.log(say)
// 输出结果为：I’ m 7 years old.
```

4　其他类型

（1）数组

在 TypeScript 中，将元素定义成数组有两种方式：第一种方式为通过在已知元素类型后添加“[]”定义一个数组；第二种方式为使用数组泛型，其实现方式是通过 Array< 元素类型 > 定义数组，示例代码如下所示。

注：数组中的元素类型均为定义数组时的元素类型。

```
// 通过在已知元素类型后添加“[]”定义一个数组
let test: number[] = [2, 7, 9];
// 使用数组泛型，其实现方式是通过 Array< 元素类型 > 定义数组
let test: Array<number> = [2, 7, 9];
```

当数组中的元素类型不同时，通过使用元组可以解决该问题。元组的主要作用是存储不同类型的元素。存储的元素要与元组中声明的数据类型顺序一致，示例代码如下所示。

```
// 声明一个元组类型
let x: [string, number];
// 初始化它
x = ['Crystal', 18];
// 当访问一个已知索引的元素，会得到正确的类型
console.log(x[0].substr(1));
// 输出结果为：rystal
```

（2）枚举（enum）

枚举类型的主要作用是作为整型常数的集合被使用，且枚举类型为集合成员赋予有意义的名称，增强可读性，示例代码如下所示。

```
enum Color {Red, Blue, Yellow};
let c: Color = Color. Blue;
```

枚举类型具有方便、快速调取数据的优点。如：通过 let colorName: string = Color[2]（调取上面代码数据），可以快速查找编号为 2 的相应数据（编号默认为从 0 开始，也可自定义赋值），示例代码如下所示。

```
// 编号默认从 0 开始，通过自定义赋值，为 Red 的编号赋值，使编号从 1 开始
enum Color {Red = 1, Blue, Yellow};
let colorName: string = Color[3];
alert(colorName);
// 弹出内容为 Yellow 的弹出窗
```

（3）任意值（Any）

当获取到不确定数据类型的变量值时，编程人员不希望类型检查器对这些变量值进行检查。为了使它们直接通过编译阶段的检查，可以使用 Any 类型对这些变量进行标记，示例代码如下所示。

```
let testAny: any = 9;
testAny= "This is Angular !";
testAny= false;
```

对于只清楚部分数据类型的数组也可使用 Any 类型，示例代码如下所示，输出结果如图 2.6 所示。

```
let test: any[] = [1, true, "free"];
// 编号默认从 0 开始，赋值给 true
test[1] = 100;
console.log(test)
```

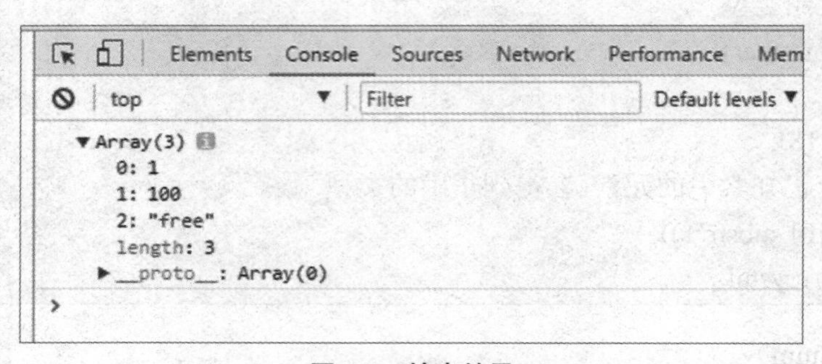

图 2.6 输出结果

快来扫一扫！

提示：对于学生最有价值的，并不是在学校读过书的事实，而是求学的态度。学，可以立志；学，可以成才；学，永远不能停止。

技能点 3　函数

1　函数定义

函数是可以从其他地方被调用执行的语句块，具有代码清晰易懂、便于维护等优点。在 TypeScript 中，可以给函数指定类型，这样在编译阶段会避免很多错误，使用户更好的对函数进行操作。与 JavaScript 相比，TypeScript 为函数添加了一些新的功能，如剩余函数、重载等。使用 TypeScript 可以创建多种类型的函数，其主要有无参函数、有参函数和匿名函数，示例代码如下所示。

```
// 无参函数
function 函数名称 () {
// 函数体
}
// 有参函数，x、y 为参数
function 函数名称 (x, y) {
    return x+y;
}
// 匿名函数
// let 变量名称 =function( 参数 1, 参数 2){}
let myAdd = function(x, y) {
return x+y;
};
```

2　函数参数

（1）可选参数和默认参数

函数的参数在 JavaScript 中是可有可无的，但在 TypeScript 中，每个函数的参数都是必须存在的（可传递 null 或 undefined 作为参数），当函数被调用时，编译器会检查每个参数是否都有提供值。传递给函数的参数个数必须与函数期望的参数个数相匹配，示例代码如下所示。

```
function buildName(firstName: string, lastName: string) {
    return firstName + " " + lastName;
}
// 传递的参数个数与上面定义的参数类型个数一致
let result3 = buildName("Bob", "Adams");
```

```
console.log(result3)
// 结果为：Bob Adams
```

当不确定传入某一类型参数时，可通过在参数后面加入“?”符号，使该参数变为可选参数，示例代码如下所示。

```
function buildName(firstName: string, lastName?: string) {
    if (lastName){
        return firstName + " " + lastName; }
    else{
        return firstName; }
}
let result1 = buildName("Bob");
let result3 = buildName("Bob", "Adams");
console.log(result1);
// 结果为：Bob
console.log(result3);
// 结果为：Bob Adams
```

在 TypeScript 中，当用户未提供参数时，可根据需要为该参数设置默认值，示例代码如下所示。

```
function buildName(firstName: string, lastName = "Smith") {
    return firstName + " " + lastName;
}
let result1 = buildName("Bob");
console.log(result1)
// 结果为：Bob Smith
```

（2）剩余参数

当多个参数同时存在时，可以创建一个变量（省略号（...）后面为变量名），将同一类型参数包裹起来，最终形成剩余函数。剩余函数与可选参数类似，可以存在（任意个参数同时存在），也可不存在（一个参数也没有）。示例代码如下所示。

```
function buildName(firstName: string, ...restOfName: string[]) {
  return firstName + " " + restOfName.join(" ");
}
let buildNameFun: (fname: string, ...rest: string[]) => string = buildName;
```

（3）this 参数

在使用 TypeScript 开发过程中，经常会用到“this”。在使用 this 时，需要创建一个方法，使参数对象绑定到 this 上。示例代码如下所示。

```
var people={
    name:[" Dream","Backlight","Autism","Tearl"],
    getName:function(){
        return function(){
        // Math.floor(Math.random() *4) 对数值进行向下取整
            var i=Math.floor(Math.random()*4);
            return {
                n:this.name[i]
            }
        }
    }
}
var pname=people.getName();
alert(" 名字："+pname().n);
// 在弹出框中弹出：名字：undefined
```

在运行期间，会发现调取不到 name 属性值，原因是 getName 中的 this 关键字指向的是 getName，访问不到外部的 name 属性，所以，需要使用箭头函数，使 this 指向参数对象，示例代码如下所示。

```
var people = {
    name:[" Dream","Backlight","Autism","Tearl"],
    getName: function () {
        return ()=> {
            var i = Math.floor(Math.random() * 4);
            return {
                n: this.name[i]
            }
        }
    }
}
var pname = people.getName();
alert(" 名字：" + pname().n);
// 在弹出框中弹出：名字：Backlight
```

（4）重载

重载是指在同一个类中，函数或方法名称相同，其参数类型、类型顺序、参数个数不同，且与返回值无关。目的是用相同的方法名实现类似的功能。在 TypeScript 中，重载的主要作用是根据传入不同的参数返回不同类型的数据，方便函数的调用与返回，代码如下所示。

```
let suits = ["hearts", "spades", "clubs", "diamonds"];
// 传入一个 suit，它把对应的编号（或 ID）返回出来
function pickCard(x: {suit: string; card: number; }[]): number;
// 传入一个编号，把对应的 suit 返回出来
function pickCard(x: number): {suit: string; card: number; };
// 具体实现方法，主要检查参数 x
function pickCard(x): any {
    // 检查我们处理的是什么类型的参数
    // 如果是数组对象，则给定 deck 并且选择 card
    if (typeof x == "object") {
        // Math.floor(Math.random() * x.length) 对数值进行向下取整
        let pickedCard = Math.floor(Math.random() * x.length);
        return pickedCard;
    }
    // 否则只选择 card
    else if (typeof x == "number") {
        let pickedSuit = Math.floor(x / 13);
        return { suit: suits[pickedSuit], card: x % 13 };
    }
}
let myDeck = [{ suit: "diamonds", card: 2 }, { suit: "spades", card: 10 }, { suit: "hearts", card:
4 }];
let pickedCard1 = myDeck[pickCard(myDeck)];
alert("card: " + pickedCard1.card + " of " + pickedCard1.suit);
// 在弹出框中弹出：card: 10 of spades
let pickedCard2 = pickCard(15);
alert("card: " + pickedCard2.card + " of " + pickedCard2.suit);
// 在弹出框中弹出：card: 2 of spades
```

技能点 4 类

1 属性

在面向对象（类的实例）方法中，表示对象相关特征的数据称为对象的属性，其具有提高

工作效率等优点。如 project 类中可以包含 id、name 和 age 等属性。类中的每个属性均包含一个可选的类型。如 id 和 age 可定义为 Number 类型，name 可定义为 String 类型，示例代码如下所示。

```
class project{
    id: number;
    name: string;
    age: number;
}
```

2　方法

要想调用方法，首先要有该对象的实例（可通过 new 创建实例），然后在类对象实例的上下文中可以定义方法。在方法中，通过“this. 属性名”可实现访问类中的某个属性（this 关键字），示例代码如下所示。

```
class project{
    id:number;
    name:string;
    age: number;
    greet() {
        console.log("Hello", this.name);
    }
}
// 调用 greet 方法之前，需要创建 project 类的实例对象
var p : project;
p = new project();
p.name = 'Angular';
p.greet();
// 输出结果为 Hello Angular
```

3　构造函数

构造函数（constructor）是在类进行实例化时执行的特殊函数。一个类中可具备多个构造函数，通过其参数个数的不同来区分。默认情况下，创建的是无参构造函数，示例代码如下所示。

```
class project {
    constructor(); {}
}
var p=new project ();
```

当需要带参构造函数时，需手动创建。通过带参构造函数可以将对象的创建工作参数化，示例代码如下所示。

```
class project {
    id: number;
    name: string;
    age: number;
    constructor(id: number, name: string, age: number) {
        this.id =id;
        this.name = name;
        this.age = age;
    }
    greet() {
        console.log("My name is "+this.name+",I am "+this.age+" years old"+".My number
            is "+this.id);
    }
}
var p: project = new project(1, 'Angular', 36);
p.greet();
// 输出结果为 My name is Angular,I am 36 years old.My number is 1
```

4 继承

继承是面向对象的重要特性之一，在 TypeScript 中作为核心语法来使用。其主要作用是子类通过继承（用 extends 关键字实现）获取父类的行为，最终，可在子类中重写、修改以及添加行为，创建父类（project 类）示例代码如下所示。

```
class project{
    constructor(name:string) {
        this.name=name;
    }
    name:string;
    sayName():void{
        console.log(this.name);
  }
}
```

创建子类（Extendsproject 类）继承父类（project 类）行为，示例代码如下所示，运行效果如图 2.7 所示。

```
class Extendsproject extends project {
    // 继承了父类的 name，添加了 success
    constructor(name:string,success:string) {
        super(name);
        this.success=success;
    }
    success:string;
    sayJob():void{
        console.log(this.success);
    }
}
var subClass=new SubClass('Linda','Engineer');
subClass.sayJob();
subClass.sayName();
console.log(subClass);
```

```
Engineer                                                    test.js:28
Linda                                                       test.js:16
▼SubClass {name: "Linda", success: "Engineer"}              test.js:35
    name: "Linda"
    success: "Engineer"
  ▶__proto__: BaseClass
>
```

图 2.7　结果

快来扫一扫!

提示：学会了 TypeScript 中的类之后，或许还不理解类中公有、私有修饰符的应用。扫描图中二维码，获取更多信息！

通过下面八个步骤的操作，实现图 2.2 所示的智慧工厂主界面的效果。

第一步：将主界面分为头部导航部分、左侧导航栏部分、轮播图以及主体内容部分和底部版权信息组件。

第二步：创建头部导航、左侧导航栏、轮播图、主体内容以及底部版权信息等组件。

第三步：设置头部内容。使用 Bootstrap 设计布局与样式，代码如 CORE0201 所示，设置样式前效果如图 2.8 所示。

代码 CORE0201：头部内容

```
<div class="navbar navbar-inverse navbar-fixed-top">
 <div class="navbar-inner">
   <div class="container-fluid">
     <!-- 智慧工厂标题 -->
     <div class="sidebar-toggle-box hidden-phone">
       <img src="../../../assets/login_icon.png">
     </div>
     <a class="dropdown-toggle " >
        智慧工厂中央管理系统
     </a>
    <!-- 右侧导航栏 -->
     <ul class=" pull-right top-menu" >
       <li class="dropdown mtop5">
         <a class="dropdown-toggle element" data-placement="bottom" data-toggle
            ="tooltip" data-original-title="Chat">
          <i class="fa fa-comments" ></i>
        </a>
      </li>
      <li class="dropdown mtop5">
       <a class="dropdown-toggle element" data-placement="bottom" data-toggle
           ="tooltip" data-original-title="Help">
          <i class="fa fa-headphones" ></i>
       </a>
      </li>
      <li class="dropdown">
       <a class="dropdown-toggle" data-toggle="dropdown">
         <span class="username" >Jhon Doe</span>
       </a>
      </li>
    </ul>
  </div>
```

```
  </div>
</div>
```

图 2.8　设置样式前

设置头部样式，为其设置背景颜色，修改字体图标位置、大小和颜色等。部分代码如 CORE0202 所示，设置样式后效果如图 2.9 所示。

代码 CORE0202：设置样式

```
.navbar-inner li.dropdown .dropdown-toggle i {
   font-size: 20px;
   color: white
}
.navbar-inverse .navbar-inner {
   background-color: #3983c2;
   background-repeat: repeat-x;
   border-color: #3983c2;
   color: #fff;
}
.sidebar-toggle-box {
   float: left;
   height: 60px;
   margin-left: -20px;
```

```
    margin-right: 20px;
    padding: 0 30px;
}
li{
    top: 10px;
    display:inline;
    padding-right: 30px;
}
.sidebar-toggle-box{
    padding-top: 10px
}
.dropdown-toggle{
    color: white;
    font-size: 24px;
    line-height: 60px
}
.pull-right{
    padding-top: 10px;
    list-style-type:none;
}
```

图 2.9　设置样式后

第四步：在 app.component.html 文件中对各个组件进行布局，设置组件的渲染位置，并设计左侧导航菜单，代码如 CORE0203 所示，设置样式前效果如图 2.10 所示。

图 2.10　设置样式前

代码 CORE0203：左侧导航

```
<!-- 头部部分 -->
<div>
  <app-navbar></app-navbar>
</div>
<div>
 <div class="row" "id="container">
  <div class="col-lg-3 sidebar-scroll" >
   <aside id="sidebar"class="sidebar row nav-collapse collapse" >
    <ul id="nav" class="sidebar-menu">
    <li class="active sub-menu">
     <a >
       <i class="fa fa-home" ></i>
       首页 <i class="fa fa-caret-right "></i>
      </a>
    </li>
    <li class="has_sub sub-menu">
```

```
        <a >
          <i class="fa fa-user "></i>
          人员档案 <i class="fa fa-caret-right "></i>
        </a>
      </li>
      <li class="has_sub sub-menu">
        <a >
          <i class="fa fa-table"></i>
          能源管理 <i class="fa fa-caret-right "></i>
        </a>
      </li>
      <li class="has_sub sub-menu" >
        <a >
        <i class="fa fa-file-audio-o"></i>
        水监控管理 <i class="fa fa-caret-right "></i>
      </a>
    </li>
    <li class="has_sub sub-menu">
      <a >
        <i class="fa fa-file-audio-o"></i>
        气报表管理 <i class="fa fa-caret-right "></i>
      </a>
    </li>
    <li class="has_sub sub-menu">
      <a >
        <i class="fa fa-bar-chart"></i>
        环安管理 <i class="fa fa-caret-right "></i>
      </a>
    </li>
    <li class="has_sub sub-menu">
      <a >
        <i class="fa fa-gear"></i>
        权限管理 <i class="fa fa-caret-right "></i>
      </a>
    </li>
  </ul>
</aside>
```

```
    </div>
   <div class="col-lg-9" id="main-content">
        <router-outlet></router-outlet>
     </div>
  </div>
</div>
<!-- 底部 -->
<div >
  <app-footer></app-footer>
</div>
```

设置左侧导航栏组件样式，实现鼠标移动到列表项上时改变背景颜色。部分代码如CORE0204所示，设置样式后效果如图2.11所示。

代码 CORE0204：设置样式

```
.sidebar{
  width: 230px;
  float: left;
  display: block;
  background:#f2f2f2;
  color: #777;
  position: relative;
}
.sidebar .sidebar-dropdown{
  display: none;
}
.sidebar ul{
  padding: 0px;
  margin: 0px;
}
.sidebar ul li{
  list-style-type: none;
}
.sidebar #nav {
  display: block;
  width:100%;
  margin:0 auto;
  position: absolute;
```

```
  z-index: 60;
}
.sidebar #nav li i{
  display:inline-block;
  margin-right: 5px ;
  background: #eee ;
  color:#888;
  width: 38px;
  height: 38px;
  line-height: 38px;
  text-align: center;
  border-radius: 30px;
  border: 1px solid #ccc;
}
.sidebar  #nav li span i{
  margin: 0px;
  color: #999;
  background: transparent !important;
  border: 0px;
}
.sidebar #nav > li > a:hover i, .sidebar #nav > li > a.open i, .sidebar #nav > li > a.subdrop i{
  color: #fff;
    background-color: #167cac !important;
    border: 1px solid #167cac;
}
.sidebar #nav > li > a:hover span i, .sidebar #nav > li > a.open span i, .sidebar #nav > li > a.
ubdrop span i{
  color: #fff;
  background: transparent !important;
  border: 0px;
}
.sidebar #nav li ul { display: none; background: #efefef url("../img/cream.png") repeat; }
.sidebar #nav li ul li a {
    display: block;
    background: none;
    padding: 10px 0px;
    padding-left: 30px;
    text-decoration: none;
```

```
    color: #777;
    border-bottom: 1px solid #ddd;
    box-shadow: inset 0px 1px 0px rgba(255, 255, 255, 0.1);
}
.sidebar #nav li ul li a:hover {
    background: #eee;
    border-bottom: 1px solid #ddd;
}
```

图 2.11　设置样式后

实现左侧菜单点击某一选项后呈选中状态，引入的 JS 代码如 CORE0205 所示。

代码 CORE0205：引入 JS 代码

```
$(document).ready(function(){
    $(window).resize(function()
    {
// 对屏幕分辨率变换时进行判断
if($(window).width() >= 765){
            $(".sidebar #nav").slideDown(350);
        }
        else{
```

```
            $(".sidebar #nav").slideUp(350);
        }
    });
    $("#nav > li > a").on('click',function(e){
        if($(this).parent().hasClass("has_sub")) {
            e.preventDefault();
        }
        if(!$(this).hasClass("subdrop")) {
            // 隐藏任何打开的菜单并删除所有其他内容
            $("#nav li ul").slideUp(350);
            $("#nav li a").removeClass("subdrop");
        // 打开菜单并显示所对应的内容
            $(this).next("ul").slideDown(350);
            $(this).addClass("subdrop");
        }
        else if($(this).hasClass("subdrop")) {
            $(this).removeClass("subdrop");
            $(this).next("ul").slideUp(350);
        }
    });
});
```

第五步：在 app.module.ts 文件中配置路由，代码如 CORE0206 所示。

代码 CORE0206：配置路由

```
import {HomepageModule}from'./homepage/homepage.module'
import {HomepageComponent} from "./homepage/homepage/homepage.component";
import { RouterModule, Routes } from '@angular/router';
//…
const appRoutes: Routes = [
//...
  { path: 'dash', component: HomepageComponent },
 ];
@NgModule({
//…
imports: [
  RouterModule.forRoot(appRoutes),
],
```

```
//…
})
export class AppModule { }
```

在 app.component.html 文件中使用路由，代码如下所示。

```
<li class="active sub-menu">
  <a routerLink="/dash">
    <i class="fa fa-home" ></i>
    首页 <i style="float: right" class="fa fa-caret-right "></i>
  </a>
</li>
```

第六步：设置轮播图，轮播图组件由轮播的图片和轮播导航等组成，通过 Bootstrap 设置其样式。代码如 CORE0207 所示。轮播图效果如图 2.12 所示。

代码 CORE0207：轮播图组件

```
<div class="carousel slide" data-ride="carousel">
  <ol class="carousel-indicators"
    <li class="active"></li>
    <li></li>
    <li></li>
  </ol>
  <div class="carousel-inner">
    <div class="item active">
      <img src="../../../../assets/welcome.jpg" alt="" class="slide-image" >
    </div>
      <!-- 部分代码省略（轮播图片）-->
  </div>
  <a href="javascript:$('.carousel').carousel('prev')" class="left carousel-control">
    <span class="glyphicon glyphicon-chevron-left"></span>
  </a>
  <a href="javascript:$('.carousel').carousel('prev')" class="right carousel-con
trol">
    <span class="glyphicon glyphicon-chevron-right"></span>
  </a>
</div>
```

图 2.12 轮播图组件

第七步：设置主体内容组件，数据显示采用 NgFor 指令遍历循环。代码如 CORE0208 所示，效果如图 2.13 所示。

代码 CORE0208：主体内容

```
<div class="mainright " >
  <div class="contain1 ">
    <div class="container-fluid">
      <div class="row"  *ngFor="let product of products">
          <!-- 企业文化 -->
          <div class="col-sm-6">
          <h2>
            企业文化 Corporate Culture
          </h2>
               <!-- 核心价值观 -->
          <div class="box box-solid">
            <div class="box-header with-border">
              <h3 > 核心价值观 Company Core Value</h3>
            </div>
            <div class="box-body clearfix">
            <div class="pull-left text-center valueinfo" >
              <div class="solid" ></div>
```

```
                <div class="title">Commitment</div>
            </div>
            <div class="pull-left text-center valueinfo" >
              <div class="title"> 诚信 </div>
              <div class="solid"></div>
              <div class="title">Integrity</div>
            </div>
            <div class="pull-left text-center valueinfo">
              <div class="title"> 创新 </div>
              <div class="solid"></div>
              <div class="title">Innovation</div>
            </div>
            <div class="pull-left text-center valueinfo" >
              <div class="title"> 进取 </div>
              <div class="solid"></div>
              <div class="title">Aggressive</div>
            </div>
        </div>
        </div>
          <!-- 使命 -->
        <div class="box box-solid">
          <div class="box-header with-border">
            <h3> 使命 Mission</h3>
          </div>
          <div class="box-body">
            <blockquote>
              <p>{{product.mission}}</p>
              <p>{{product.mission1}}</p>
            </blockquote>
          </div>
      </div>
      <!-- 愿景 -->
      <div class="box box-solid">
        <div class="box-header with-border">
          <h3> 愿景 Vision</h3>
        </div>
        <div class="box-body">
          <blockquote>
```

```
                    <p>{{product.vision}}</p>
                    <p>{{product.vision1}}</p>
                </blockquote>
            </div>
            <!-- 理念 -->
        </div>
        <div class="box box-solid">
            <div class="box-header with-border">
                <h3> 理念 Idea</h3>
            </div>
            <div class="box-body">
                <blockquote>
                    <p>{{product.idea}}</p>
                    <p>{{product.idea1}}</p>
                </blockquote
            </div>
        </div>
    <!-- 省略部分代码(管理方针)-->
      </div>
    </div>
   </div>
  </div>
</div>
```

对应界面的 ts 文件,定义一个类并编写其属性,之后声明一个变量用于接收数据。代码如 CORE0209 所示。

代码 CORE0209:对应的 ts 代码

```
import { Component, OnInit } from '@angular/core';
@Component({
  selector: 'app-product',
  templateUrl: './product.component.html',
  styleUrls: ['./product.component.css']
})
export class ProductComponent implements OnInit {
  // 定义一个变量
  private  products: Array<Product>;
  constructor() { }
```

```
  // 定义一个数组
  ngOnInit() {
    this.products=[
      new Product(
        " 为半导体、太阳能、光通讯等高新技术产业和国防工业提供高性能的产品和服务，实现中国石英的崛起。",
        " 做世界一流的高新材料供应商，打造百年菲利华品牌。",
        " 对外全面满足客户需求，对内最大限度满足员工物质和自我价值实现的要求，切实履行企业公民责任。",
        " 以不断的技术进步和严格的管理提高产品品质和服务的同时，注重环保、节能的社会责任，始终坚持健康、持续的发展方向。",
        " 以遵章守法为基础，坚持“安全第一、预防为主”的原则，持续改进职业健康安全工作，创建和谐企业。",
        "Providing high quality products and  services for high-tech industry such as Semiconductor,solar energy,optical communication etc.and national defense indusrty to achieve the rise of Chinese quartz.",
        "To be the world class adcanced Hi-tech material supplier and build a century Felihua brand.",
        "Externally  satisfying the customers' demands and internally meet the request of physical needs as well as realization of staff's self-value，essentially fulfilling the responsibility of enterprise citizenship.",
        "We Keep on improving the quality of products &amp; services by constant technical progress and strict management, and in the meanwhile take our social responsibilities on environmental protection and energy saying aiming at healthy and sustainable development.",
        "Being disciplined &amp; law-abiding,we devote on improving the conditions for occupational safety and health and a harmonious enterprise wht the principle of 'Safety first,precaution crucial'."
      )
    ]
  }
}
// 定义一个类
export class Product{
  constructor(
    public mission:string,
    public vision:string,
    public idea:string,
    public policy:string,
```

```
        public health:string,
        public mission1:string,
        public vision1:string,
        public idea1:string,
        public policy1:string,
        public health1:string
    ){
    }
}
```

图 2.13 内容组件

第八步：底部为版权信息制作，利用 <p> 标签即可完成制作，代码如 CORE0210 所示。

代码 CORE0210：底部组件

```
<div class="container" >
  <footer>
    <div class="row">
    <div class="col-lg-12">
      <p>@Angular</p>
    </div>
  </div>
```

```
    </footer>
</div>
```

至此，智慧工厂主界面模块制作完成。

本项目通过对智慧工厂主界面模块的学习，对 TypeScript 的基本知识具有初步了解，学会如何使用内置类型在编译时执行类型检查。能够使用函数定义执行内容的位置，了解 TypeScript 类的属性及方法等。为以后编写 Angular 项目打好基础。

Tuple	元组
enum	枚举
Boolean	布尔类型
void	空值
constructor	构造函数
extends	继承
sentence	判断
undefined	不明确的
substr	字符串的子串

一、选择题

1.TypeScript 内置类型不包括以下哪种（　）。

（A）Number　（B）Text　（C）Boolean　（D）Void

2.TypeScript 特点正确的是（　）。

（A）JavaScript 不可以使用 TypeScript　（B）TypeScript 不支持其他 JS 库

（C）TypeScript 不能转化为 JS　（D）TypeScript 具有便携性

3. TypeScript 中函数的主要作用不包含以下哪种（　）。

（A）抽象层　（B）模拟类　（C）信息隐藏　（D）组件

4. 对 TypeScript 使用类的方法说法正确的是（　）。

（A）方法是运行在类对象实例上下文的函数

（B）方法是运行在类对象实例的函数

（C）方法是运行在类对象实例上文的函数

（D）方法是运行在类对象实例下文的函数

5. 在 TypeScript 里，对继承说法错误的是（　）。

（A）继承是 TypeScript 中的核心语法

（B）子类可以从父类获得父类的行为

（C）子类中可重写、修改以及添加行为

（D）子类中不可对从父类继承的行为进行修改

二、填空题

1.TypeScript 是 ________ 的超集。

2.TypeScript 中包含的任何类型的数字（无论是整数或者浮点）均属于 ______ 类型，均用 ______ 表示。

3.TypeScript 的枚举类型属于 _______ 类型。

4. 构造函数的作用是 ____。

5.TypeScript 通过属性可以 ______________________________。

三、上机题

使用 Angular 编写符合以下要求的页面。

要求：用 Angular 创建两个组件实现如图效果。

card works!

card works!
卡头
我是卡的内容

项目三　智慧工厂人员档案模块

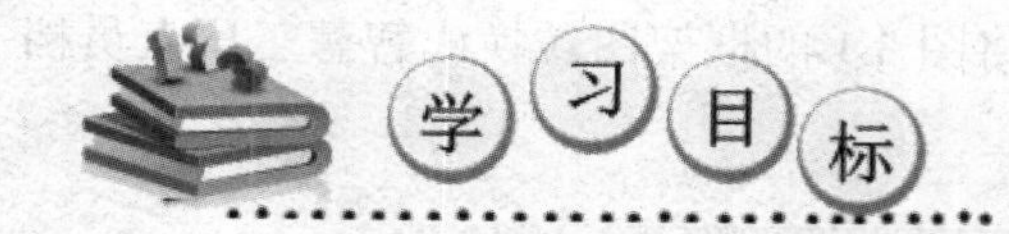

通过智慧工厂人员档案模块功能的实现，了解人员档案模块中数据显示所需知识，学习 Angular 架构、模板语法等相关知识，掌握使用 NgFor 指令读取数据，具有解决使用 Angular 数据显示时可能出现问题的能力。在任务实现过程中：

- 了解人员档案模块的数据显示。
- 学习 Angular 的模板语法。
- 掌握 Angular 的 NgFor 指令。
- 具有解决数据显示问题的能力。

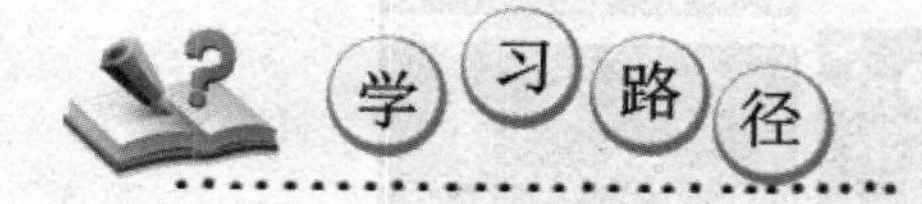

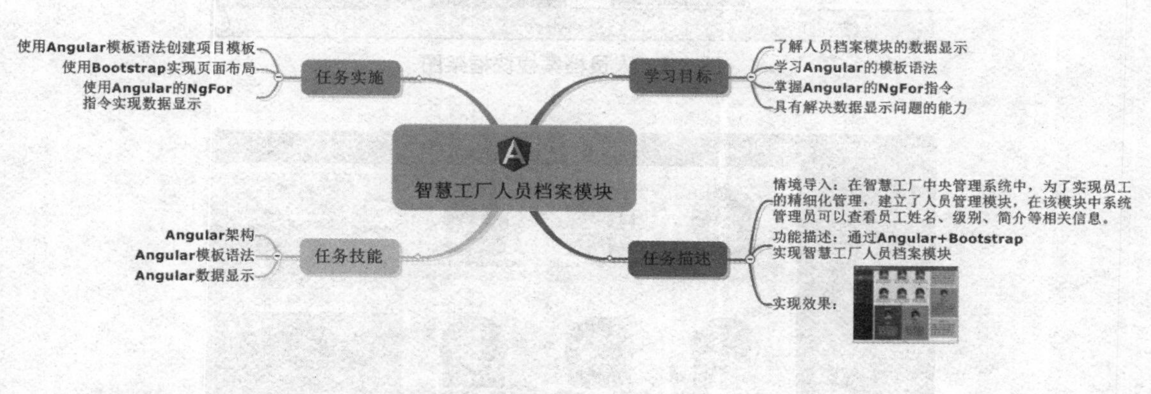

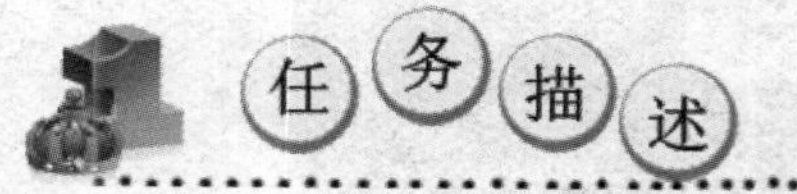

【情境导入】

在智慧工厂中央管理系统中，为了实现员工的精细化管理，建立了人员档案模块，在该模块中系统管理员可以查看员工姓名、级别、简介等相关信息。同时，系统管理员也可将优秀员工的信息以轮播图的形式显示。本项目主要是通过实现智慧工厂的人员档案模块来学习

Angular 的模板语法和数据显示。

【功能描述】

使用 Bootstrap+Angular 实现智慧工厂人员档案模块：

- 使用 Angular 模板语法创建项目模板。
- 使用 Bootstrap 实现页面布局。
- 使用 Angular 的 NgFor 指令实现数据显示。

【基本框架】

基本框架如图 3.1 所示，通过本项目的学习，能将图 3.1 的框架图转换成智慧工厂人员档案模块，效果图如图 3.2 所示。

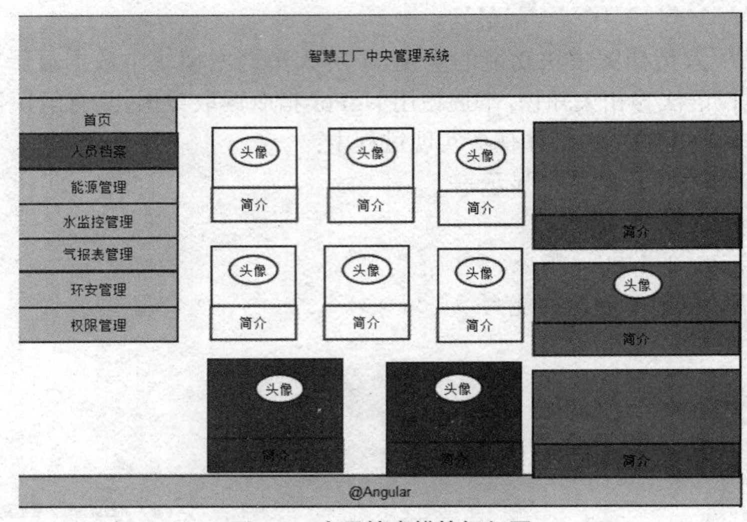

图 3.1　人员档案模块框架图

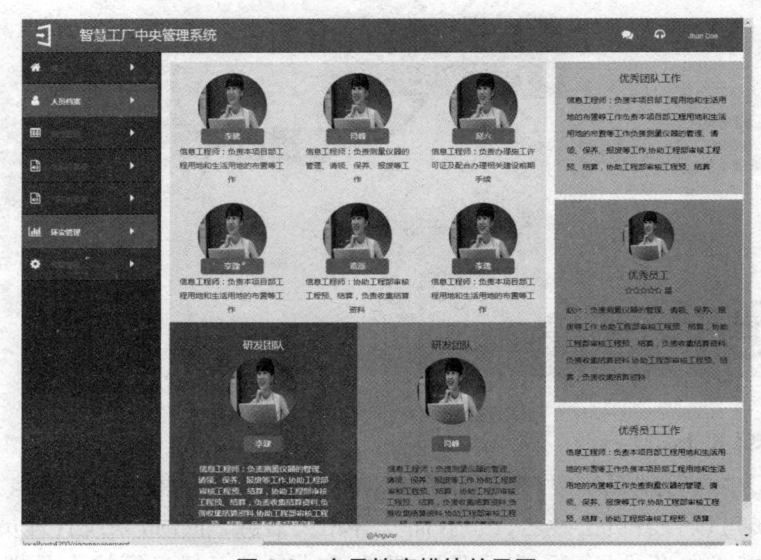

图 3.2　人员档案模块效果图

技能点 1　Angular 架构

Angular 是一款比较完善的前端框架，具有模板功能强大、自带指令丰富等优点，比较适用于大型项目，如电商应用或手机 App 应用等。Angular 可以在一个页面上构建多种功能，通过 TypeScript 可以实现页面交互效果，主要包括模块、组件、模板、元数据、数据绑定、指令、服务、依赖注入等。使用 Angular 框架，可以节省开发时间，提高效率。其结构如图 3.3 所示。

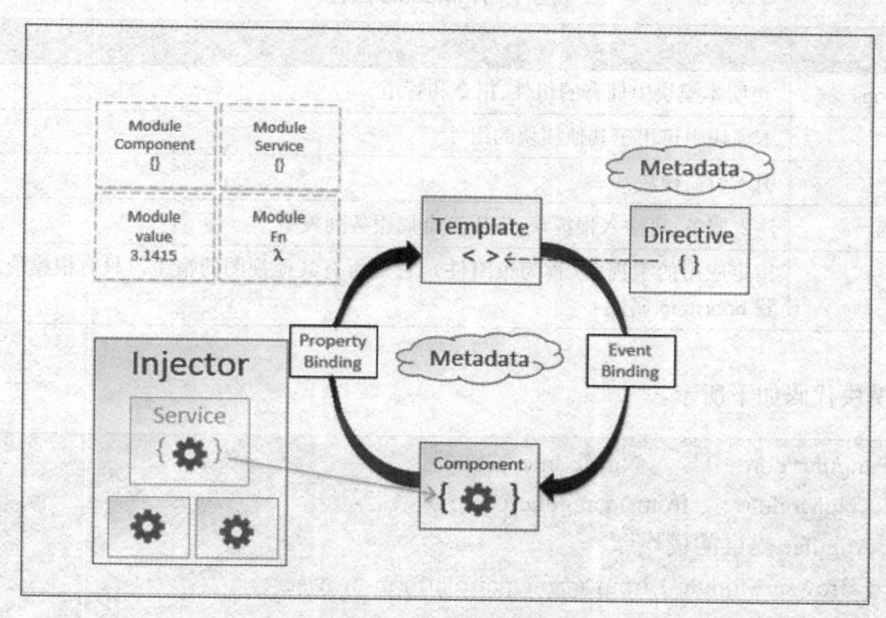

图 3.3　Angular 结构

使用 Angular 开发应用模块时，首先创建模块，在模块中创建组件与服务，然后在组件中编写 HTML 模板（可使用指令、数据绑定与用户交互），之后注入服务添加应用逻辑，最后通过引导根模块启动应用。

（1）模块

Angular 应用是模块化的，每个应用中至少含有一个模块，即根模块。模块的主要作用是对组件与服务等进行打包，形成内聚的功能块。通过在命令窗口输入“ng g module 模块名称”即可创建模块。其在应用启动时主动加载，也可以由路由进行异步惰性加载。Angular 提供了多种模块库，每种库的前缀名都带有 @angular，通过命令 npm install 即可安装，使用 import 语句进行导入。模块结构如图 3.4 所示。

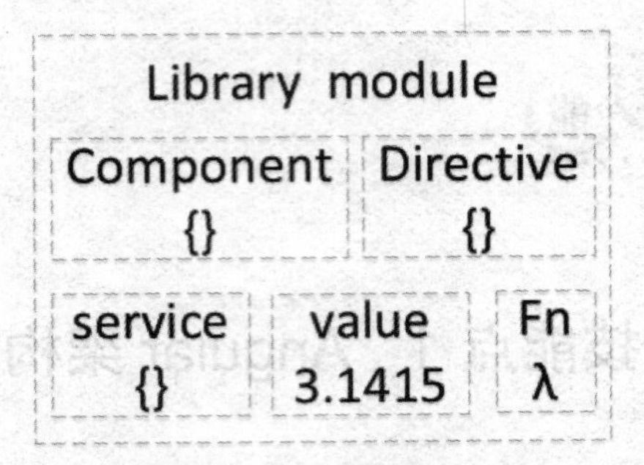

图 3.4 模板结构

Angular 模块是带有 @NgModule（装饰器）的类，主要作用是接收元数据对象。其属性如表 3.1 所示。

表 3.1 NgModule 属性

属性	描述
declarations	声明本模块中拥有的组件、指令和管道
exports	本模块中可用于其他模块的组件
imports	引入所需模块
providers	注入服务，可注入根模块，应用到全局服务列表中
bootstrap	指定应用的主视图（称为根组件），它是所有其他视图的宿主。只有根模块才能设置 bootstrap 属性

使用模块代码如下所示。

```
// 从 @angular/core 中导入 NgModule 装饰器
import { NgModule }    from '@angular/core';
// 引入 Angular 提供的模块库
import { BrowserModule } from '@angular/platform-browser';
@NgModule({
// 引入相关的模块，此处只有一个模块
   imports:        [ BrowserModule ],
   providers:    [ AppService],
   declarations: [ AppComponent ],
   exports:          [ AppComponent ],
// 使应用一开始会加载 AppComponent 组件到 index 页面中
   bootstrap:       [ AppComponent ]
})
```

（2）组件

界面中的视图区域是由组件组成的，通过在组件中编写 HTML 模板，最终在界面中被渲染。在组件的 ts 文件中还可定义应用逻辑等。在 Angular 中，应用程序可包含多个组件，方便程序设计者进行编程。

使用组件时，首先创建一个模块（或直接在根模块内），在模块内通过命令窗口输入“ng g component 组件名称”命令创建组件，创建成功后，生成名为“组件名称”的文件夹（包含 .html、.css、.ts、.spec.ts（测试代码）文件），并在所在模块中会自动导入组件，如：创建 project 模块，在该模块内通过命令创建 project 组件，project.module.ts 代码如下所示。

```
import { ProjectComponent } from './project/ project .component';
//…
@NgModule({
//...
declarations: [ProjectComponent]
})
export class ProjectModule { }
```

注意：如果是复制过来的组件，一定要在组件所在模块下进行如上配置。

（3）模板

Angular 模板一般以 HTML 形式存在，其可在 .html 文件中编译，也可在 .ts 文件中使用 template 属性进行编译。通过组件自身所含有的模板可以定义组件视图。其主要作用是定义如何渲染组件。在模板中可以使用指令、数据绑定等功能，除此之外还可以在模板中使用点击事件、定义数组和显示自定义组件等（自定义组件能和原生的 HTML 无缝融合）。

（4）元数据

在创建组件后，对应的 ts 文件中会自动生成一个 @Component 装饰器，其包含多个配置项，这些配置项的值即为元数据。元数据的主要作用是渲染组件并执行组件的逻辑。@Component 装饰器配置项如表 3.2 所示。

表 3.2　配置项

配置项	描述
selector	使 Angular 在父级 HTML 中查找相对应的标签，创建并插入该组件
templateUrl	组件 HTML 模板的模块相对地址
styleUrls	对应的 CSS 文件相对地址
providers	组件所需服务的依赖注入

如：通过命令创建 project 组件，在 project.component.ts 文件中，@Component 中的元数据示例代码如下所示。

```
@Component({
  selector: 'app-project',
  templateUrl: './project.component.html',
```

```
  styleUrls: ['./project.component.css'],
  // 在注入服务时使用(项目五具有详细讲解)
  providers:[]
})
export class ProjectComponent implements OnInit {
}
```

(5)数据绑定

数据绑定可以使模板和组件之间相互关联,相互合作。通过在模板 HTML 中绑定标记,可以实现数据绑定。数据绑定的语法具有多种形式,其形式如表 3.3 所示。

表 3.3　绑定形式

形式	描述	形式
插值表达式	传递的是值;方向是组件→模板;单项绑定	<h1>{{title}}</h1>
属性表达式	传递的是属性;父组件→子组件;单项绑定	<img [src]="imgUrl" />
事件绑定	组件控制器的一个方法绑定到模板元素的事件上,处理点击等事件	<button (click)="show()"></button>
双向绑定	组件和模板保持同步。无论组件和模板哪一方改变,另一方都同步改变	<input [(XX)]="hero.age">

双向绑定是数据绑定中最重要的语法之一,如:通过表单双向绑定,可以将绑定后的属性值通过属性绑定传递给输入框,用户在输入框中进行修改并通过事件绑定再次传递给组件,最后对属性值进行刷新。使用数据绑定示例代码如下所示。

```
<!-- 插值表达式 显示组件的 title 属性的值 -->
<h1>{{title}}</h1>
<!-- 属性表达式:显示将父组件的 imgUrl 的值传到子组件的 src 属性中 -->
<img [src]="imgUrl" />
<!-- 事件绑定:点击按钮时会调用组件的 show() 方法 -->
<button (click)="show()"></button>
<!-- 双向绑定 -->
<input [(ngModel)]="hero.age">
```

(6)指令

指令可以为模板元素添加一些新的功能或特性,在命令窗口输入"ng g directive 指令名称"即可创建指令文件。Angular 中指令具有三种类型:组件(拥有模板的指令)、结构型指令(用来改变 DOM 结构,是通过在 DOM 中添加、移除和替换元素来修改布局的指令)、属性型指令(以元素属性形式来使用的指令,用来修改一个现有元素的外观或行为,并实现双向绑定)。使用结构型指令和属性型指令示例代码如下所示。

```
<!-- 结构型指令 -->
<li *ngFor="let hero of heroes"></li>
<hero-detail *ngIf="selectedHero"></hero-detail>
<!-- 属性型指令 -->
<!-- ngModel 实现了数据双向绑定 -->
<input [(ngModel)]="hero.name">
```

（7）服务

服务在应用中是非常重要的，可以是类，也可以是对象或者方法。其实现目的是当重载或刷新页面时，数据不会被清除，而且与加载之前保持一致。Angular 具有自己的一系列项目结构，在项目中通过在命令窗口输入"ng g service 服务名称"即可创建服务，其主要用来在特性模块或者应用中共享数据和方法。将服务注入到最高层的组件或模块中，其子组件或子模块也可应用。

（8）依赖注入

通过依赖注入可以提供类的新实例，并负责处理类所需的全部依赖，其主要作用是将服务注入到需要的模块、组件或服务中。使用 Angular 实现依赖注入时，首先需要注入器提供令牌（依赖的标识）来获取服务，通常在构造函数里面，为参数指定类型，该参数类型就是依赖注入器所需的令牌；Angular 把该令牌传给提供器，提供器根据该令牌创建被依赖的对象。

技能点 2　Angular 模板语法

模板的主要作用是将组件内容在页面中显示，作为视图所用，Angular 模板具有多种语法，如 HTML 语法、表达式、绑定语法、模板引用变量、模板表达式操作符等。

1　HTML 语法

模板的 HTML 语法丰富多样，与 Angular 结合使用克服了其在构建应用上的不足，如通过指令等可以扩展模板中的 HTML 元素。Angular 模板语法中兼容大部分 HTML 元素，不兼容 HTML 的元素有：<script>、<html>、<body> 和 <base> 等。使用 HTML 语法示例代码如下所示。

```
<h1> Angular 模板的 HTML</h1>
<p>The sum of 1 + 1 is {{1 + 1}}</p>
 <b>Total:</b> {{qty * cost | currency}}
```

2　模板表达式

在编程代码中，双花括号 {{…}} 中的内容即为模板表达式。模板表达式的主要作用是进

行运算并显示其结果。其相比于 JavaScript 具有新的模板运算符：管道操作符（|）、安全导航操作符（?.）、非空断言操作符（!）。模板表达式对应用是非常重要的，它具有以下特点：

- 没有副作用：模板表达式除了目标属性的值以外，不改变应用的任何状态。
- 执行迅速：Angular 执行模板表达式非常的频繁，每一次按键或鼠标移动后都会被调用。
- 应用简单：常规方法是属性名或方法调用。
- 幂等性：总是返回完全相同的东西，直到某个依赖值发生改变。

注：模板表达式只能引用表达式的上下文（模版中各种绑定值的来源）。

使用模板表达式示例代码如下所示。

```
{{title}}
<span [hidden]="isUnchanged">changed</span>
<div *ngFor="let hero of heroes">{{hero.name}}</div>
<input #heroInput> {{heroInput.value}}
```

3 绑定语法

通过绑定语法可以协调用户所见视图和应用数据（数据源）交互。具有代码编写简单、易阅读和维护等优点。Angular 提供多种数据绑定，其根据数据流可分为三种：从数据源到视图、从视图到数据源、从视图到数据源再到视图。数据绑定分类如表 3.4 所示。

表 3.4 数据绑定分类

数据方向	语法
从数据源到视图目标	{{expression}} [target]="expression" bind-target="expression"
从视图目标到数据源	(target)="statement" on-target="statement"
从视图到数据源再到视图	[(target)]="expression" bindon-target="expression"

通过绑定语法可以进行绑定目标，该目标可以是 property、事件、attribute 等，目标汇总如表 3.5 所示。

表 3.5 目标汇总

目标	语法
（元素 | 组件 | 指令的）property	<img [src]="heroImageUrl"> <hero-detail [hero]="currentHero"></hero-detail> <div [ngClass]="{special: isSpecial}"></div>
（元素 | 组件 | 指令的）事件	<button (click)="onSave()">Save</button> <hero-detail (deleteRequest)="deleteHero()"></hero-detail> <div (myClick)="clicked=$event" clickable>click me</div>

续表

目标	语法
事件与 property	<input [(ngModel)]="name">
attribute（例外情况）	<button [attr.aria-label]="help">help</button>
class property	<div [class.special]="isSpecial">Special</div>
style property	<button [style.color]="isSpecial ? 'red' : 'green' ">

（1）属性绑定

使用属性绑定可以实现将属性设置成模板表达式。属性绑定为单向绑定，方向是从组件到元素。属性绑定的基本语法为 [属性名]，其中属性名的作用是标记目标属性。属性绑定具有多种绑定形式，如表 3.6 所示。

表 3.6　绑定语法分类

分类	语法
元素属性设置为组件属性的值	<img [src]="heroImageUrl">
设置指令的属性	<div [ngClass]="classes">[ngClass] binding to the classes property</div>
设置自定义组件的模型属性（这是父子组件之间通讯的重要途径）	<hero-detail [hero]="currentHero"></hero-detail>

使用属性绑定示例代码如下所示。

```
<p><span [innerHTML]="title"></span> is the <i>property bound</i> title.</p>
<p><img [src]="heroImageUrl"> is the <i>property bound</i> image.</p>
```

（2）事件绑定

在项目设置过程中，通常会用事件绑定监听处理应用逻辑，如在登录等界面，用户会在输入框等表单中输入信息，当信息确认时，会点击按钮进行发送等操作，通过事件绑定可监听这些操作。事件绑定也为单向绑定，但数据流是从元素传到组件上。事件绑定语法由等号左侧带圆括号的目标事件和右侧引号中的模板语句组成。示例代码如下所示。

```
<!-- 圆括号中的名称标记出目标事件 -->
<button (click)="onSave()">Save</button>
<!--on- 前缀的备选形式 -->
<button on-click="onSave()">On Save</button>
<!--myClick 需要匹配上已知指令的事件属性 -->
<div (myClick)="clickMessage=$event" clickable>click with myClick</div>
```

在使用事件绑定时，也可设置事件处理器，如当触发事件时，处理器会根据 $event 的事件对象传递信息来执行模板语句（在 HTML 中获取值并存入模型中），示例代码如下所示。

```
<!-- 要更新 name 属性，就要通过路径 $event.target.value 来获取更改后的值。-->
<input [value]="user.name" (input)="user.name=$event.target.value" >
```

（3）双向绑定

双向绑定的主要目的是方便用户实时更改信息，使用双向绑定既可以做到设置元素属性，又可以监听元素事件变化。双向绑定语法格式为 [(xxx)]，它是由事件绑定 (xxx) 与属性绑定 [xxx] 的结合。使用双向绑定示例如下所示。

```
<div>
<button (click)="dec()" title="smaller">-</button>
<button (click)="inc()" title="bigger">+</button>
<!-- 点击按钮，在最小 / 最大值范围限制内增加或者减少 size。最终 size 触发 sizeChange 事件。-->
<label [style.font-size.px]="size">FontSize: {{size}}px</label>
</div>
```

对应 ts 文件示例代码如下所示。

```
import { Component, EventEmitter, Input, Output } from '@angular/core';
@Component({
  selector: 'app-product',
  templateUrl: './product.component.html',
  styleUrls: ['./product.component.css'],
})
export class AppComponent {
//@Input 是用来定义模块的输入，size 的初始值是一个输入值，来自属性绑定
@Input()  size: number | string;
//@Output 是用来定义模块的输出
 @Output() sizeChange = new EventEmitter<number>();
 dec() { this.resize(-1); }
 inc() { this.resize(+1); }
 resize(delta: number) {
   this.size = Math.min(40, Math.max(8, +this.size + delta));
   this.sizeChange.emit(this.size);
 }
}
```

4　模板引用变量

模板引用变量的主要作用是用来引用模板中的某个 DOM 元素、指令等，其常用的语法为使用“#”声明变量（或用 ref- 前缀代替 #）。它具有完全独立、不用与 Component 进行交互等优点。这些变量可以提供在模块中直接访问元素的能力，如声明在 <input> 上的变量在模板另一侧的 <button> 上使用。模板引用变量示例代码如下所示。

```
<!-- phone 引用了 input 元素，通过点击事件将 value 传递给事件 -->
<input #phone>
<button (click)="callPhone(phone.value)">Call</button>
<!-- fax 引用了 input 元素，并将 value 传递给事件句柄（ref- 前缀代替 #）-->
<input ref-fax >
<button (click)="callFax(fax.value)">Fax</button>
```

5　模板表达式操作符

（1）管道操作符（|）

管道是一个接收输入值并返回转换结果的函数，具有多种转换形式，如英文字母大小写切换、数字转换为金额等。其实现方式是通过管道操作符（|）将左侧的表达式结果传给右侧的管道函数进行转换。使用管道操作符示例代码如下所示。

```
<div>
  {{title | uppercase }}
</div>
<!--json 管道调试绑定 -->
<div>{{currentHero | json}}</div>
```

（2）安全导航操作符 (?.)

安全导航操作符的主要作用是用来保护出现在属性路径中的 null 和 undefined 值，具有流畅、便利的优势。实现效果为当表达式遇到第一个空值时跳出，可以正常工作。使用安全导航操作符示例代码如下所示。

```
<!-- 当 currentHero 为空时，保护视图渲染器，让它免于失败 -->
The current hero’s name is {{currentHero?.name}}
The title is {{title}}
```

（3）非空断言操作符（!）

当使用 --strictNullChecks 切换到新的严格空值检查模式时，在该模式下，具有类型的变量默认是不允许 null 或 undefined 值的，如果出现 null 或 undefined 值时，类型检查器将会抛出一个错误。但是，有时我们需要类型检查器对特定的属性表达式，不做严格空值检测，这时需要非空断言操作符来要告诉类型检查器，它不会为空，非空断言操作符不会防止出现 null 或

undefined。使用非空断言操作符示例代码如下所示。

```
<div *ngIf="hero">
The hero’ s name is {{hero!.name}}
</div>
```

提示：在 Angular 中，模板具有两种使用方式。

第一种：在 .html（如 app.component.html）文件中使用模板，代码如 CORE0301 所示。

```
CORE0301：app.component.html
<div>
  Angular!
</div>
```

在对应的 .ts（app.component.ts）文件中获取模板，代码如 CORE0302 所示。

```
CORE0302：app.component.ts
import { Component } from '@angular/core';
@Component({
  selector: 'app-root',
  // 通过 templateUrl 获取模板数据
  templateUrl: './app.component.html',
  styleUrls: ['./app.component.css']
})
export class AppComponent {
}
```

第二种：在 .ts（app.component.ts）文件中直接使用，代码如下所示。

```
import { Component } from '@angular/core';
@Component({
  selector: 'app-root',
// 通过 templateUrl 获取模板数据
  template: `
    <div>
        Angular!
    </div>
  `,
 styleUrls: ['./app.component.css']
})
```

```
export class AppComponent {
}
```

提示：当我们对 Angular 模板语法了解后，你是否意识到工作中团队合作的重要性呢？扫描图中二维码，了解关于团队合作的幽默小故事。

技能点 3　Angular 数据显示

1　插值表达式

数据显示的最基本方式是使用插值表达式绑定组件的属性名。其语法结构是：{{+(组件属性名(表达式))+}}，实现方式是 Angular 对所有插值表达式进行求值，并将结果与相邻字符串连接起来，最终把组合结果赋值给元素或指令，渲染到视图模板上，使用插值表达式示例代码如下所示。

```
<p>My name is {{name}}</p>
<!-- 计算后的字符串进行赋值 -->
<img src="{{heroImageUrl}}" style="height:30px">
<!--Angular 先对它求值，再把它转换成字符串 -->
<p>The sum of 1 + 1 is {{1 + 1}}</p>
<!-- 这个表达式可以调用组件的方法 -->
<p>The sum of 1 + 1 is not {{1 + 1 + getVal()}}</p>
```

2　NgFor

NgFor 的主要作用是可以循环从数组中取值并显示出来，类似于 for 循环。在 Angular 中，数据经常以数组的形式显示，通过使用 NgFor 指令可以遍历循环这些数据，最终使用表达式渲染到视图层。要循环输出一组数组，首先需要设置属性，并为其添加相应的数据，然后使用 NgFor 遍历循环该数据，并通过表达式渲染到视图层。使用 NgFor 指令效果如图 3.5 所示。

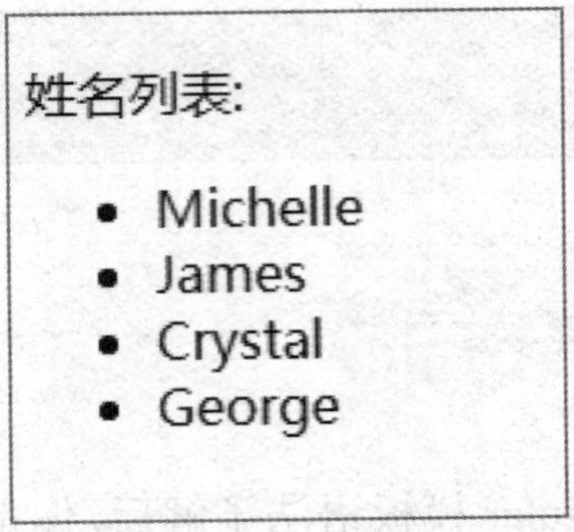

姓名列表:

- Michelle
- James
- Crystal
- George

图 3.5 NgFor 指令

为了实现图 3.5 效果,代码如 CORE0303 所示。

代码 CORE0303: app.component.html

```
<p>{{title}}:</p>
<ul>
  <li *ngFor="let name of nameList">
    {{ name}}
  </li>
</ul>
```

对应 ts 代码如 CORE0304 所示。

代码 CORE0304: app.component.ts

```
import { Component } from '@angular/core';
@Component({
  selector: 'app-root',
  templateUrl: './app.component.html',
  styleUrls: ['./app.component.css'],
})
export class AppComponent {
  title = ' 姓名列表 ';
  nameList = ['Michelle', 'James', 'Crystal', 'George']
}
```

3 NgIf

NgIf 的主要作用是根据提供的条件决定是否显示或隐藏这个元素,使用时需将 NgIf 绑定到元素上。在项目中,经常使用 NgIf 根据表达式的值,将 then(默认是 NgIf 指令关联的内联模板)或 else(默认是 null)模板的内容渲染到指定位置,指令语法如表 3.7 所示。

表 3.7　指令语法

语法	形式
基本语法	<div *ngIf="condition">...</div> <ng-template [ngIf]="condition"><div>...</div></ng-template>
else 语法	<div *ngIf="condition; else elseBlock">...</div> <ng-template #elseBlock>...</ng-template>
then 和 else 语法	<div *ngIf="condition; then thenBlock else elseBlock"></div> <ng-template #thenBlock>...</ng-template> <ng-template #elseBlock>...</ng-template>
as 语法	<div *ngIf="condition as value; else elseBlock">{{value}}</div> <ng-template #elseBlock>...</ng-template>

使用 NgIf 实现点击 show 按钮，显示内容，点击 hide 按钮，出现内容隐藏时显示的文本，示例代码如下所示，效果如图 3.6 所示

show　内容隐藏时显示的文本

hide　显示内容

图 3.6　NgIf 的使用

```
import { Component,OnInit } from '@angular/core';
@Component({
  selector: 'app-root',
  template: `
      <button (click)="show = !show">{{show ? 'hide' : 'show'}}</button>
      <div *ngIf="show; then thenBlock; else elseBlock"></div>
      <ng-template #thenBlock> 显示内容 </ng-template>
      <ng-template #elseBlock> 内容隐藏时显示的文本 </ng-template>
`,
  styleUrls: ['./app.component.css']
})
export class AppComponent implements OnInit{
  ngOnInit() {
  }
}
```

4　NgSwitch

NgSwitch 的主要作用是从多个元素中根据 switch 条件来选取某一个（或多个）。其实现

方式是需将 NgSwitch 指令绑定到可以返回任何类型的表达式中。Angular 的 NgSwitch 是一组相互合作的指令，包括 ngSwitch、ngSwitchCase 和 ngSwitchDefault 等表现形式。具体如表 3.8 所示。

表 3.8 NgSwitch 指令表现形式

指令	描述
ngSwitch	绑定到一个返回候选值的表达式
ngSwitchCase	绑定到一个返回匹配值的表达式，当符合条件后就会显示对应的内容
ngSwitchDefault	可选，会在没有任何一个 ngSwitchCase 被选中时把它所在的元素加入 DOM 中

使用 NgSwitch 示例代码如下所示。

```
import { Component } from '@angular/core';
@Component({
  selector: 'app-root',
  template: `
    <div>
  <h1>{{title}}</h1>
  <div>
    <!-- num 为变量名 -->
    <div [ngSwitch]="num">
      <!-- A、B、C 为可选变量名，ngSwitchCase 指令描述已知结果 -->
      <div *ngSwitchCase="0">ngSwitchCase A</div>
      <div *ngSwitchCase="1">ngSwitchCase B</div>
      <div *ngSwitchCase="2">ngSwitchCase C</div>
      <div *ngSwitchCase="3">ngSwitchCase D</div>
      <!-- ngSwitchDefault 指令处理所有其他未知情况 -->
      <div *ngSwitchDefault>ngSwitchCase Default</div>
    </div>
  </div>
  <button (click)="changeElement()">Swicth</button>
</div>
 `,
  styleUrls: ['./app.component.css']
})
export class AppComponent {
  private num = 0;
  changeElement(): void {
```

```
        if (this.num > 3) {
            this.num = 0;
        }
        this.num++;
    }
}
```

快来扫一扫！

当我们学会使用 Angular 数据显示后，是否还不知道怎么应用？扫描图中二维码，使你对数据显示拥有更全面的了解。

通过下面九个步骤的操作，实现图 3.2 所示的智慧工厂人员档案模块的效果。

第一步：将人员档案模块分为人员显示、优秀团队、团队工作、优秀员工等部分。

第二步：创建主组件和人员显示、优秀团队、团队工作、优秀员工等子组件。

第三步：在主组件 personnel-management.component.html 中对模块进行布局，设置轮播图与人员信息渲染位置。代码如 CORE0305 所示。

代码 CORE0305：主组件

```
<div>
<div class="row">
  <div class="col-md-8">
    <div >
      <app-leader></app-leader>
    </div>
    <div>
      <app-introduction></app-introduction>
    </div>
  </div>
  <div class="col-md-4">
```

```
        <div>
          <app-good></app-good>
        </div>
        <div>
          <app-groups></app-groups>
        </div>
        <div>
          <app-success></app-success>
        </div>
      </div>
</div>
</div>
```

第四步：设置人员信息，通过 NgFor 指令遍历循环数据，代码如 CORE0306 所示。设置样式前效果如图 3.7 所示。

代码 CORE0306：人员显示

```
<!-- 通过 NgFor 指令遍历循环 products 数组 -->
<div *ngFor="let hero of heroes"
      [class.selected]="hero === selectedHero" class="col-md-4 col-sm-4 col-lg-4 container" >
  <div >
<div class="imag" >
      <img src="{{hero.image}}" alt="" class="image1">
</div>
      <div class="centain_test" >
      <div >
         <h5 >{{hero.title}}</h5>
      </div>
      </div>
      <div class="caption" >
         <p>{{hero.price}}: {{hero.desc}}</p>
      </div>
      </div>
      <div class="actions" >
          <a  (click)="delete(hero); $event.stopPropagation()"><i  class="fa fa-remove "></i></a>
      </div>
</div>
```

创建 hero.ts 文件，声明数据类型，代码如 CORE0307 所示。

代码 CORE0307：数据类型

```
export class Hero {
  id:number;
  image:string;
  title:string;
  price:string;
  rating:number;
  desc:string
}
```

创建 in-memory-data.service.ts 文件，添加数据，部分代码如 CORE0308 所示。

代码 CORE0308：添加数据

```
import { InMemoryDbService } from 'angular-in-memory-web-api';
export class InMemoryDataService implements InMemoryDbService {
  createDb() {
    const heroes = [
      {
        id: 0,
        image: '../../assets/women1.jpg',
        title: ' 李建 ',
        price: ' 信息工程师 ',
        rating: '4.5',
        desc: ' 负责本项目部工程用地和生活用地的布置等工作 '
      },
      {
        id: 1,
        image: '../../assets/women1.jpg',
        title: ' 肖峰 ',
        price: ' 信息工程师 ',
        rating: '4.5',
        desc: ' 负责测量仪器的管理、请领、保养、报废等工作 '
      },
      //…
  ];
return {heroes};
  }
}
```

创建 hero.service.ts 服务文件，获取数据，并对数据进行操作，代码如 CORE0309 所示。

代码 CORE0309：获取数据

```
import { Injectable }    from '@angular/core';
import { Headers, Http } from '@angular/http';
import 'rxjs/add/operator/toPromise';
import { Hero } from './hero';
@Injectable()
export class HeroService {
  private headers = new Headers({'Content-Type': 'application/json'});
  private heroesUrl = 'api/heroes';  // api
  constructor(private http: Http) { }
  getHeroes(): Promise<Hero[]> {
      return this.http.get(this.heroesUrl)
                      .toPromise()
                      .then(response => response.json().data as Hero[])
                      .catch(this.handleError);
  }
  delete(id: number): Promise<void> {
    const url = '${this.heroesUrl}/${id}';
    return this.http.delete(url, {headers: this.headers})
      .toPromise()
      .then(() => null)
      .catch(this.handleError);
  }
  private handleError(error: any): Promise<any> {
    console.error('An error occurred', error);
    return Promise.reject(error.message || error);
  }
}
```

在组件中依赖注入服务（具体操作见项目五），获取服务数据，并对其进行操作，代码如 CORE0310 所示。

代码 CORE0310：注入服务

```
import { Component, OnInit } from '@angular/core';
import {Hero} from "../../authority-management/hero";
import {HeroService} from "../../authority-management/hero.service";
```

```
@Component({
  selector: 'app-leader',
  templateUrl: './leader.component.html',
  styleUrls: ['./leader.component.css']
})
export class LeaderComponent implements OnInit {
  heroes:Hero[];
  selectedHero: Hero;
  constructor(private heroService:HeroService) {}
  getHeroes():void {
    this.heroService
      .getHeroes()
      .then(heroes => this.heroes = heroes);
  }
  delete(hero:Hero):void {
    this.heroService
      .delete(hero.id)
      .then(() => {
        this.heroes = this.heroes.filter(h => h !== hero);
        if (this.selectedHero === hero) {
          this.selectedHero = null;
        }
      });
  }
  ngOnInit():void {
    this.getHeroes();
  }
  onSelect(hero:Hero):void {
    this.selectedHero = hero;
  }
}
```

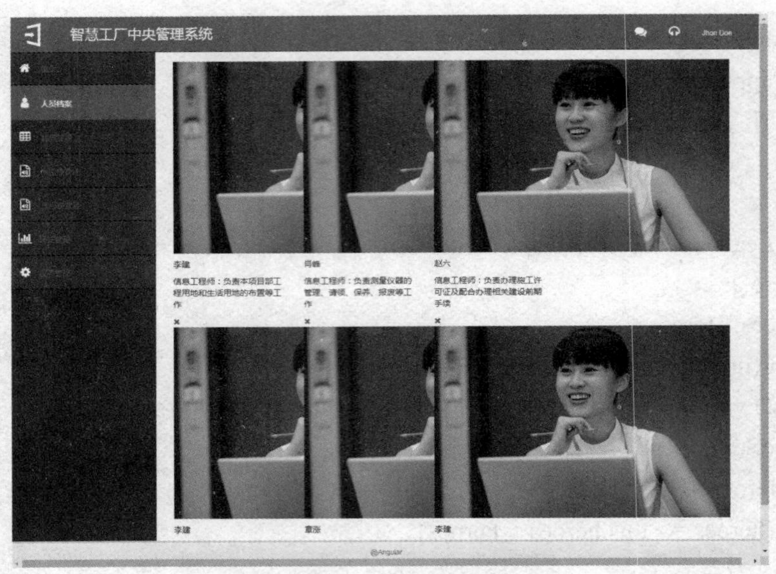

图 3.7　设置样式前

为了界面的美观，通过 border-radius 设置图片边框，并将图片居中显示。使用 CSS 定位，将姓名覆盖在图片上。设置鼠标滑动到个人信息上，删除按钮显示。代码如 CORE0311 所示。设置样式后效果如图 3.8 所示。

代码 CORE0311：CSS 样式

```
.container{
  padding-top: 20px;
  background-color: #f2f2f2
}
.image1{
  border-radius: 50%;
  height: 120px;
  width: 120px
}
.caption{
  margin-top: 10px;
  width: 99%;
  line-height: 25px;
  text-align: center
}
.imag{
  text-align: center
}
```

```
/* 名称 */
.centain_test{
    text-align: center;
    padding-left: 40px
}
.centain_test div{
    width:50%;
    background:#4a8bc2;
    top: 120px;
    position: absolute;
    color: white;
    text-align: center;
    border-radius: 5px;
}
.centain_test div h5{
    margin-top: 5px;
    margin-bottom: 5px;
}
/* 字体图标 */
.actions a{
    color: black
}
.actions a i{
    color: black
}
/* 删除隐藏 */
.actions {
    width: 64px;
    height: 32px;
    padding: 5px 8px;
    border-radius: 5px;
    position: absolute;
    top: 10%;
    left: 100%;
    margin-top: -13px;
    margin-left: -34px;
    opacity: 0;
    -moz-transition: opacity 0.3s ease-in-out;
```

```
}
.container:hover .actions {
  opacity: 1;
}
```

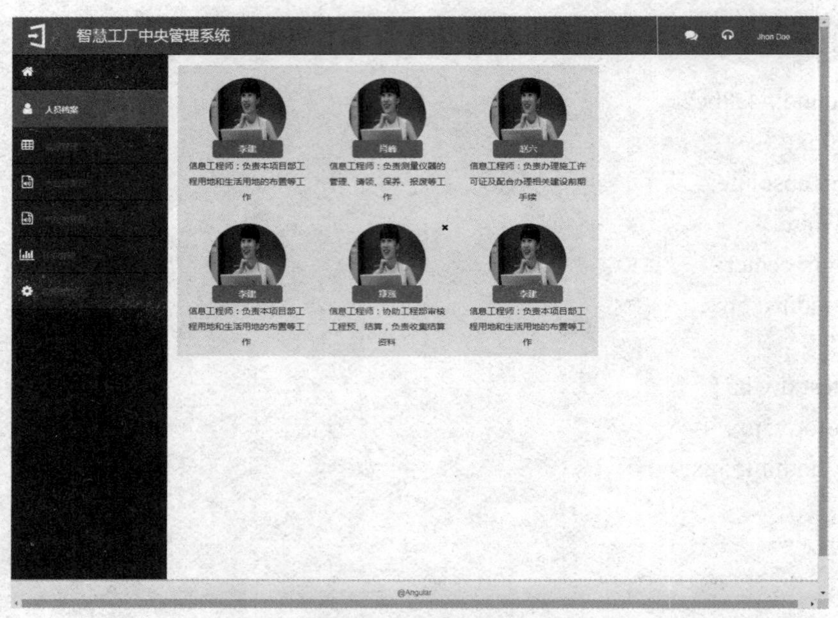

图 3.8 设置样式后

第五步：设置优秀团队，将优秀的团队成员展现出来。部分代码如 CORE0312 所示。设置样式前效果如图 3.9 所示。

代码 CORE03012：优秀员工

```
<div class="contain" >
        <div class="contain-user">
        <h4> 研发团队 </h4>
        <img src="../../../assets/women1.jpg" alt="" class="image1" >
        <div class="contain-name" >
            <h5 > 李建 </h5>
        </div>
        <div class="caption" >
            <p class="color" > 信息工程师：负责测量仪器的管理、请领、保养、报废等工作，协助工程部审核工程预、结算，协助工程部审核工程预、结算，负责收集结算资料，负责收集结算资料，协助工程部审核工程预、结算，负责收集结算资料 </p>
            </div>
```

```
        </div>
        //…
    </div>
```

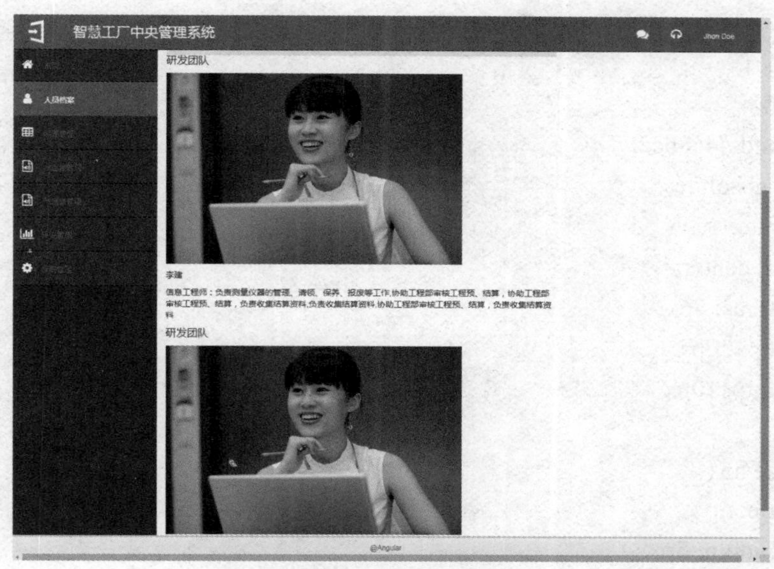

图 3.9　设置样式前

为了突出显示优秀团队的员工，设置不同的颜色背景，加以区分。将文字居中显示，使界面简洁。代码如 CORE0313 所示，设置样式后效果如图 3.10 所示。

代码 CORE0313：CSS 样式

```
.contain{
  margin-top: 20px;
}
.contain-user{
  text-align: center;
  padding: 20px;
  width: 50%;
  float: left;
  background-color: #2e6da4;
}
.contain-user h4{
   color: white
}
/* 图片 */
.image1{
```

```
    border-radius: 50%;
    height: 120px;
    width: 120px
}
/* 姓名 */
.contain-name{
     width:10%;
     background:#4a8bc2;
     position: absolute;
     color: white;
     text-align: center;
     border-radius: 5px;
     margin-top: 20px;
     margin-left: 110px
}
.contain-name h5{
     margin-top: 5px;
     margin-bottom: 5px;
}
/* 描述 */
.caption{
     margin-top: 40px;
     padding: 30px
}
.caption .color{
   color: white
}
.contain-user1{
   text-align: center;
   padding: 20px;
   width: 50%;
   float: left;
   background-color: #6fb3e0
}
```

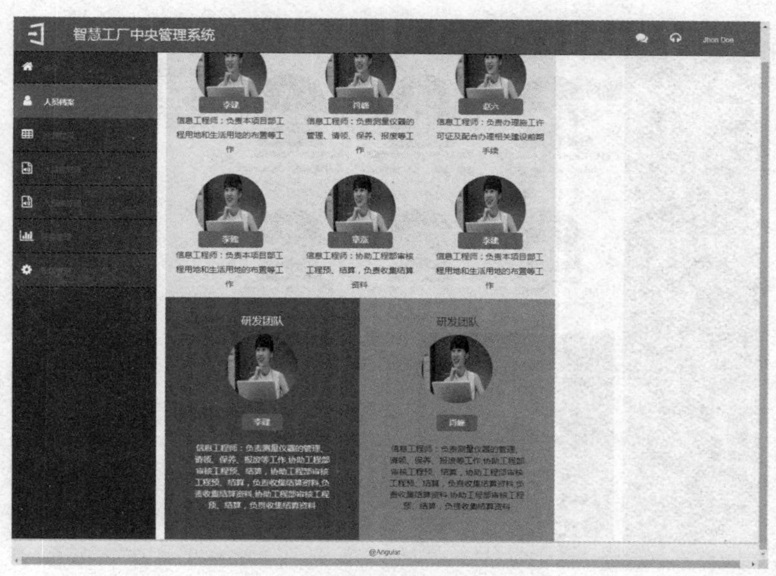

图 3.10　设置样式后

第六步：设置团队工作，将优秀团队的工作情况以文字的形式显示，增强员工之间的团结。代码如 CORE0314 所示，设置样式前效果如图 3.11 所示。

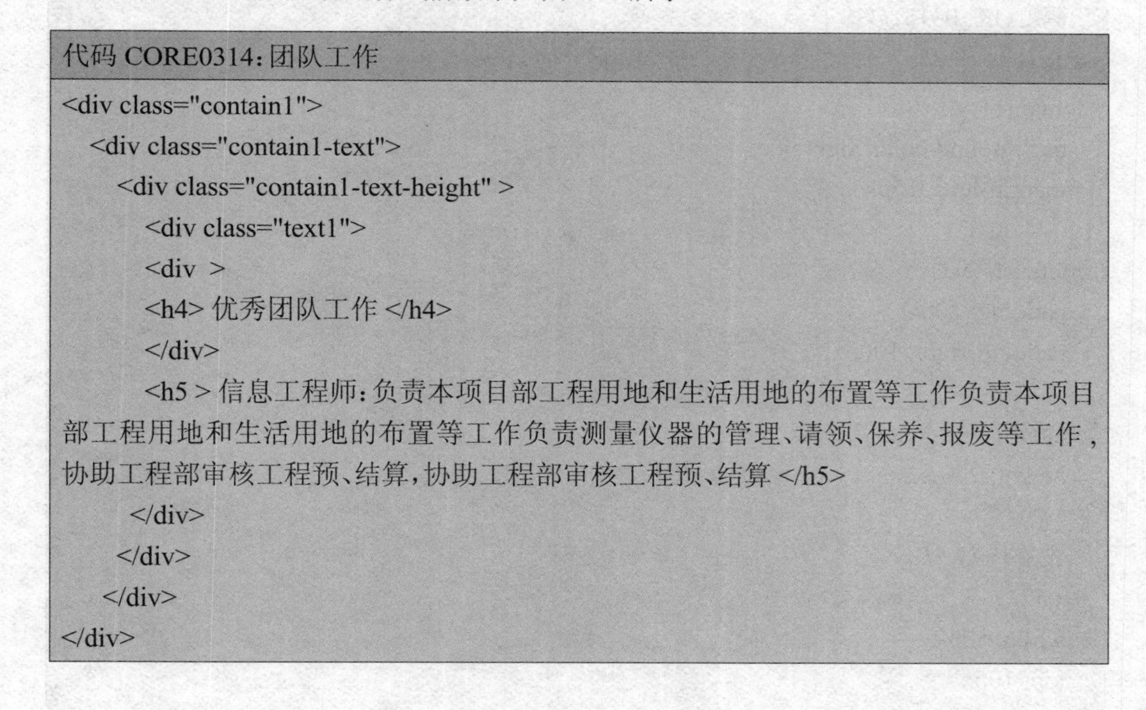

代码 CORE0314：团队工作

```
<div class="contain1">
  <div class="contain1-text">
    <div class="contain1-text-height" >
      <div class="text1">
      <div  >
      <h4> 优秀团队工作 </h4>
      </div>
      <h5 > 信息工程师：负责本项目部工程用地和生活用地的布置等工作负责本项目部工程用地和生活用地的布置等工作负责测量仪器的管理、请领、保养、报废等工作，协助工程部审核工程预、结算，协助工程部审核工程预、结算 </h5>
     </div>
    </div>
   </div>
</div>
```

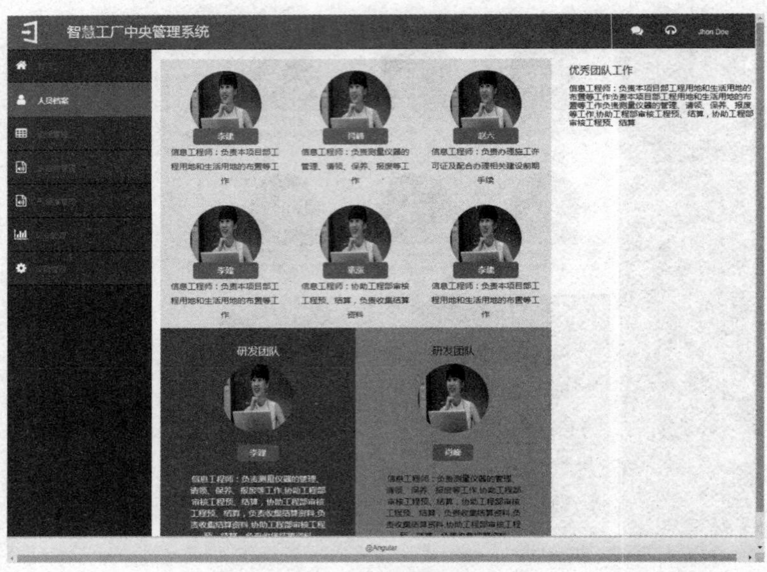

图 3.11 设置样式前

设置颜色背景，标题居中显示，代码如 CORE0315 所示，设置样式后效果如图 3.12 所示。

代码 CORE0315：CSS 样式

```
/* 整体背景 */
.contain1{
  background-color: lightblue;
  margin-left: -15px
}
.contain1-text{
  padding: 20px;
  padding-right: 100px;
}
.contain1-text-height{
  height: 200px;
}
/* 文字内容 */
.text1{
  width: 85%;
  top: 10px;
  position: absolute;
  color: black;
}
.text1 div{
```

```
    text-align: center
}
.text1 h5{
    margin-top: 5px;
    margin-bottom: 5px;
    line-height: 30px
}
```

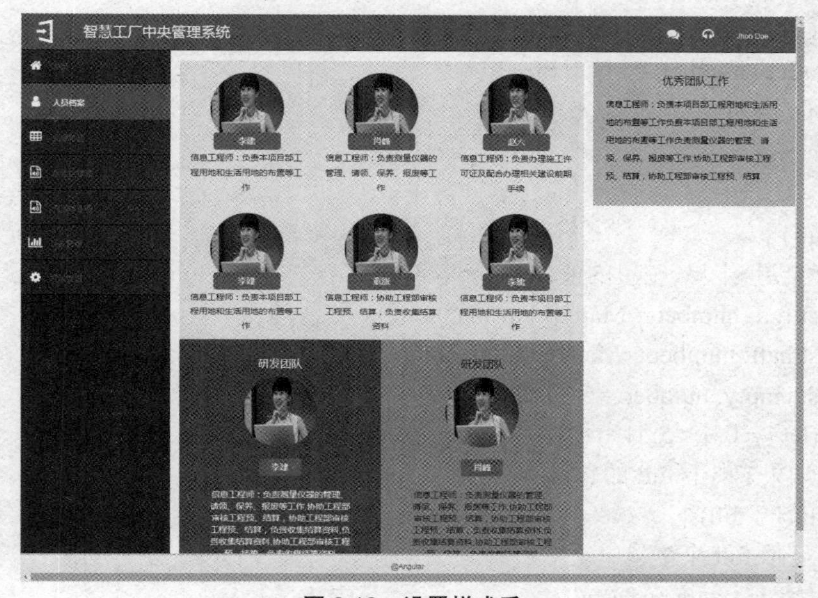

图 3.12　设置样式后

第七步：设置信用等级评分，通过信用等级评分将优秀员工显示出来，可以直观了解工程师的工作态度。代码如 CORE0316 所示。

```
代码 CORE0316：信用等级评分
<p>
    <i *ngFor="let star of stars" class="fa"
         [class.fa-star]="star==='full' "
         [class.fa-star-half-o]="star==='half' "
         [class.fa-star-o]="star==='empty' " >
    </i>
    <span>{{rating}} 星 </span>
</p>
```

对应 ts 文件代码如 CORE0317 所示。

代码 CORE0317：信用等级评分 ts 代码

```
import { Component, OnInit ,Input} from '@angular/core';
@Component({
  selector: 'app-personnelstart',
  templateUrl: './personnelstart.component.html',
  styleUrls: ['./personnelstart.component.css']
})
export class PersonnelstartComponent implements OnInit {
  @Input()
  public rating: number;
  public stars = [];
  constructor() {
  }
  ngOnInit() {
   // 定义全星、半星、空星的值，取字符串长度
      const full: number = Math.floor(this.rating);
      const half: number = Math.ceil(this.rating - full);
      const empty: number = 5 - full - half;
      for (let i = 0; i < 5; i++) {
        // 如果 i 小于 full 的长度
          if (i < full) {
        // 显示 full 个全星
              this.stars.push('full');
        // 如果 i 等于 full 的长度且 half 长度不为 0
        } else if (i === full && half !== 0) {
        // 显示 full 个全星，并显示一个半星
          this.stars.push('half')
        } else {
           this.stars.push('empty')
        }
      }
  }
}
```

第八步：设置优秀员工，通过 <img> 标签显示优秀员工图片，调用信用等级评分组件，显示其星级，并通过文字对其进行介绍。代码如 CORE0318 所示，设置样式前效果如图 3.13 所示。

代码 CORE0318:优秀员工

```
<div class="group">
<div class="container">
  <div >
    <div class="img">
      <img src="../../../assets/women1.jpg" alt="" class="image1">
    </div>
    <div class="title1">
      <h4> 优秀员工 </h4>
      <app-personnelstart></app-personnelstart>
    </div>
    <div class="caption">
      <p> 赵六:负责测量仪器的管理、请领、保养、报废等工作，协助工程部审核工程预、结算,协助工程部审核工程预、结算,负责收集结算资料，负责收集结算资料，协助工程部审核工程预、结算,负责收集结算资料 </p>
    </div>
  </div>
</div>
</div>
```

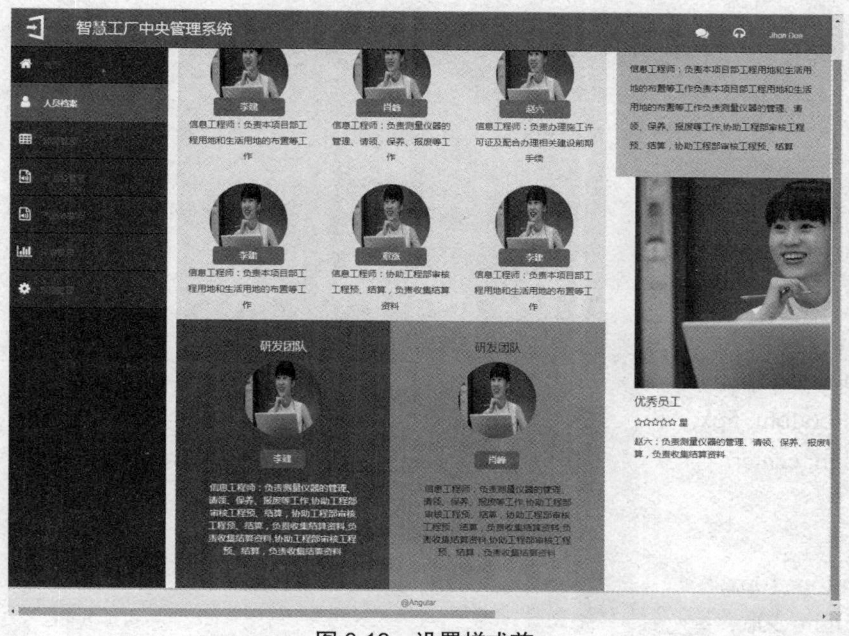

图 3.13　设置样式前

设置背景颜色，并设置标题与信用评分居中显示。代码如 CORE0319 所示，设置样式后效果如图 3.14 所示。

代码 CORE0319：CSS 样式

```
/* 布局 */
.group{
  margin-left: -15px;
  margin-top: 10px
}
.container{
  padding: 20px;
  background-color:#2ecc40;
  width: 100%
}
/* 图片样式 */
.img{
  text-align: center
}
.image1{
  border-radius: 50%;
  height: 90px;
  width: 90px
}
/* 标题样式 */
.title1{
  width: 100%;
  text-align: center
}
.title1 h4{
  margin-top: 15px;
  margin-bottom: 5px;
  text-align: center
}
.caption{
  margin-top: 10px;
  line-height: 30px
}
```

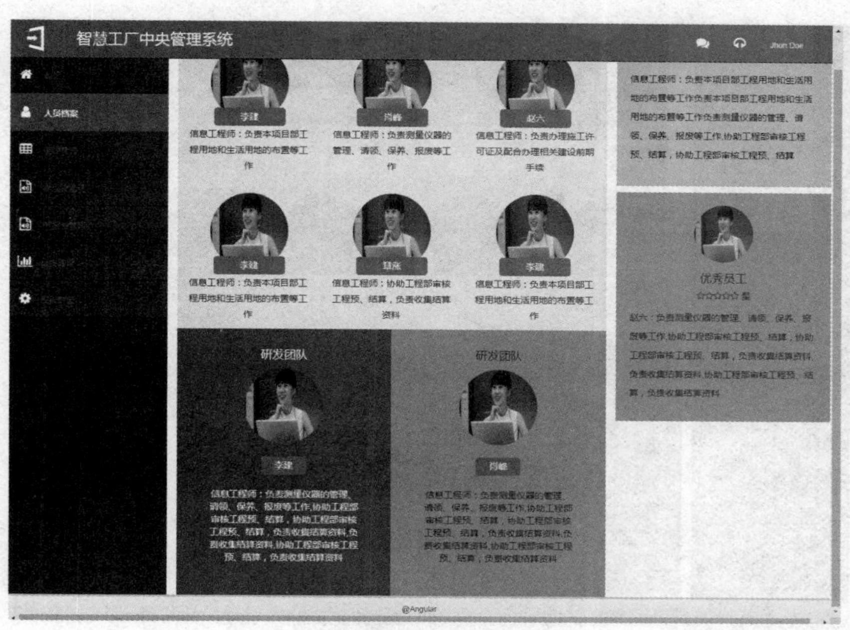

图 3.14　设置样式后

第九步：设置优秀员工工作，将优秀员工的工作情况以文字的形式显示，代码如 CORE0320 所示，设置样式前效果如图 3.15 所示。

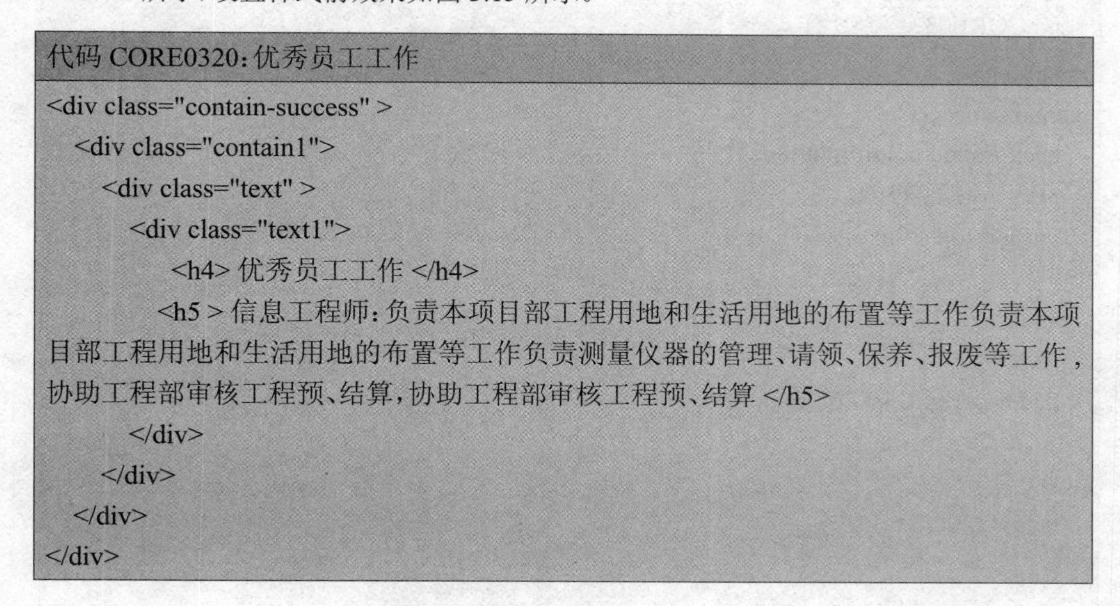

代码 CORE0320：优秀员工工作

```
<div class="contain-success" >
  <div class="contain1">
    <div class="text" >
      <div class="text1">
        <h4> 优秀员工工作 </h4>
      <h5 > 信息工程师：负责本项目部工程用地和生活用地的布置等工作负责本项目部工程用地和生活用地的布置等工作负责测量仪器的管理、请领、保养、报废等工作，协助工程部审核工程预、结算，协助工程部审核工程预、结算 </h5>
      </div>
    </div>
  </div>
</div>
```

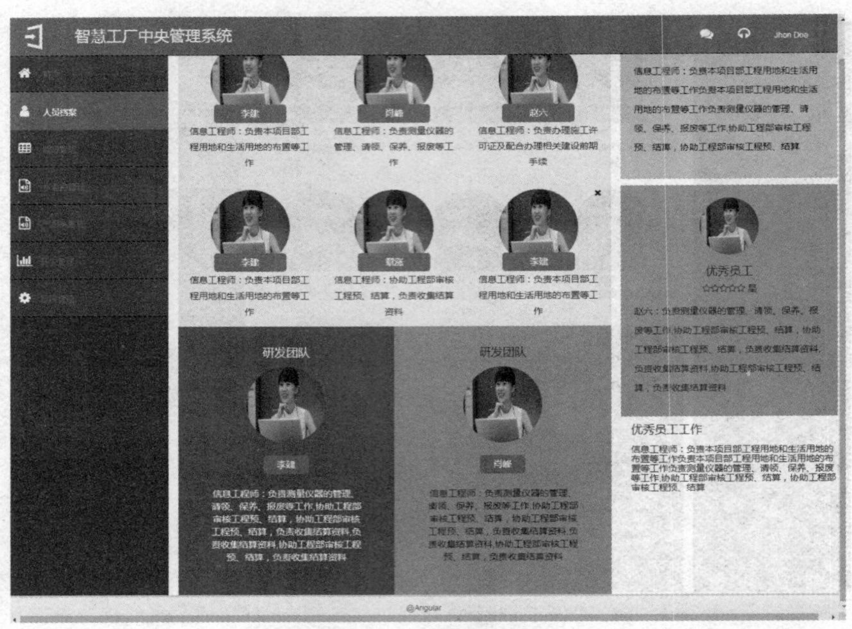

图 3.15　设置样式前

设置颜色背景，标题居中显示，代码如 CORE0321 所示，设置样式后效果如图 3.16 所示。

代码 CORE0321：CSS 样式

```
/* 布局 */
.contain-success{
   background-color: lightblue;
   margin-left: -15px;
   margin-top: 10px
}
.contain1{
   padding: 20px;
   padding-right: 100px;
}
.text{
   height: 200px;
 }
/* 文字 */
.text1{
   width: 85%;
   position: absolute;
   color: black;
```

```
}
.text1 h4{
    text-align: center
}
.text1 h5{
    margin-top: 5px;
    margin-bottom: 5px;
    line-height: 30px
}
```

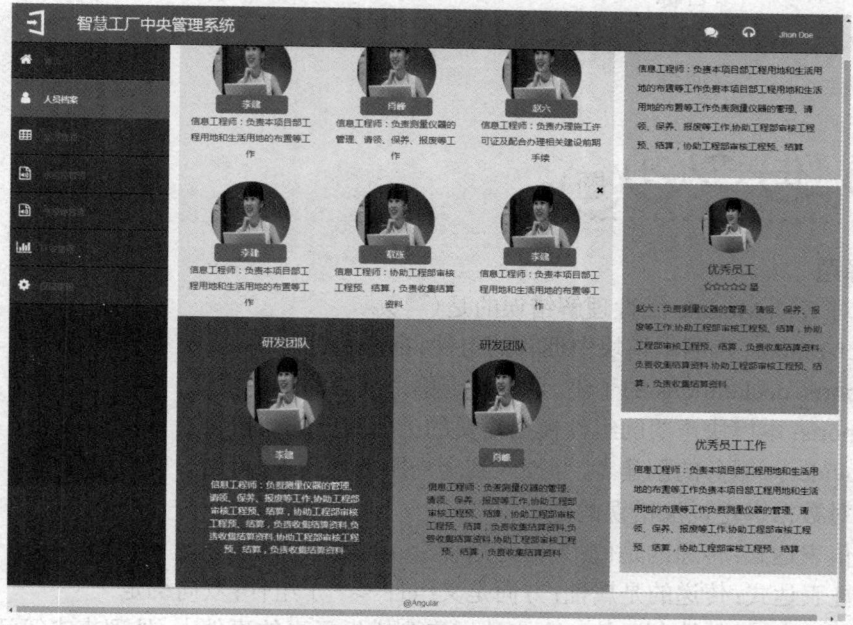

图 3.16　设置样式后

至此，智慧工厂人员档案模块制作完成。

本项目通过对智慧工厂人员档案模块的学习，对人员档案模块中显示数据功能所需知识具有初步了解，掌握了人员档案模块的页面编写、美化及功能实现的流程。具有解决使用 Angular 模板语法显示数据时可能出现问题的能力，为以后实现各个功能打好基础。

declarations　声明
exports　出口，输出
providers　提供者
imports　进口
selector　选择器
expression　表示
target　目标
property　财产
attribute　属性

一、选择题

1. 下面对 @NgModule 属性理解错误的是（　）。
（A）declarations：声明本模块中拥有的组件、指令和管道
（B）exports：declarations 的子集，只能用于单个模块的组件模板
（C）imports：本模块声明的组件模板需要的类所在的其他模块
（D）providers：服务的创建者，并加入到全局服务列表中，可用于应用任何部分

2. 下面对数据绑定说法错误的是（　）。
（A）插值表达式：传递的是值；方向是组件 -> 模板；单向绑定
（B）属性表达式：传递的是属性；方向是父组件 <-> 子组件；双向绑定
（C）事件绑定：组件控制器的一个方法绑定到模板元素的事件上，处理点击等活动
（D）双向绑定：组件和模板保持同步，无论组件和模板哪一方改变，另一方都一起同步

3. 绑定语法分类错误的是（　）。
（A）从数据源到视图目标：{{expression}}　[target]="expression"　bind-target="expression"
（B）从视图目标到数据源：(target)="statement"　on-target="statement"
（C）从视图到数据源再到视图：[(target)]="expression"　bindon-target="expression"
（D）从数据源到视图再到数据源：[target]="expression"　bind-target="expression"

4. 属性绑定说法正确的是（　）。
（A）<img src="heroImageUrl">
（B）<div ngClass="classes">[ngClass] binding to the classes property</div>
（C）<hero-detail hero="currentHero"></hero-detail>
（D）<p [ngClass]="classes"></p>

5. 指令语法书写错误的是（　）。

（A）基本语法：<div *ngIf="condition">...</div>
<ng-template [ngIf]="condition"><div>...</div><-template>
（B）else 语法：<div *ngIf="condition; else elseBlock">...</div>
<ng-template #elseBlock>...<-template>
（C）then 和 else 语法：<div *ngIf="condition; then Block else Block"></div>
<ng-template #thenBlock>...<-template>
<ng-template #elseBlock>...<-template>
（D）as 语法：<div *ngIf="condition as value; else elseBlock">{{value}}</div>
<ng-template #elseBlock>...<-template>

二、填空题

1.Angular 是一个完整的单页应用开发框架，主要是对 ______ 进行编译构建。

2.Angular 模块主要是用来将组件、指令和管道打包成内聚的功能块，还可以将 ______ 加载到应用程序中。

3. 模板表达式具有以下特点 ________、________、________、________。

4. 绑定语法的主要作用是 ________________________。

5. 使用插值表达式实现方法是 ________________________。

三、上机题

使用 Angular 编写符合以下要求的页面。

要求：使用 Angular 的 NgIf 指令 else 语法实现如图效果。

项目四　智慧工厂能源管理模块

通过智慧工厂能源管理模块的实现，了解该模块的基本布局，学习使用 Angular 组件及指令来构建应用，掌握 Angular 生命周期钩子，具有搭建 Angular 模板视图层的能力。在任务实现过程中：

- 了解智慧工厂能源管理模块的布局。
- 学习 Angular 组件、指令。
- 掌握 Angular 生命周期钩子。
- 具备搭建 Angular 模板视图层的能力。

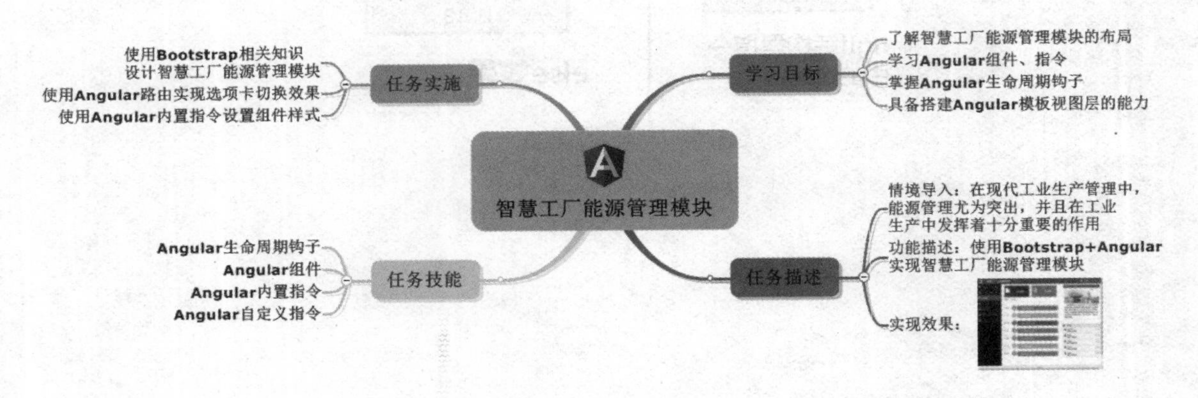

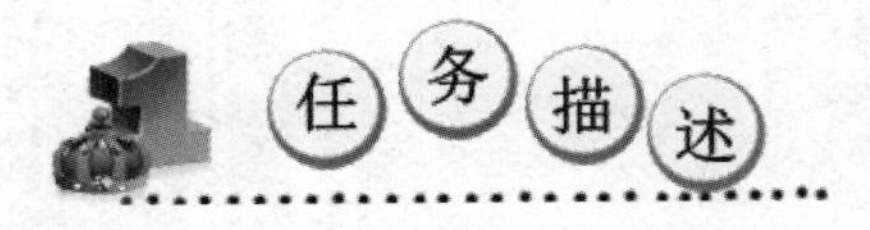

【情境导入】

在现代工业生产管理中，能源管理尤为突出，并且在工业生产中发挥着十分重要的作用。根据能源管理模块能够了解到当今热门的能源新闻，并可以对水能源和气能源进行报警监控。

本项目主要是通过实现智慧工厂的能源管理模块来学习 Angular 的组件和指令。

【功能描述】

使用 Bootstrap+Angular 实现智慧工厂能源管理模块：

- 使用 Bootstrap 相关知识设计智慧工厂能源管理模块。
- 使用 Angular 路由实现选项卡切换效果。
- 使用 Angular 内置指令设置组件样式。

【基本框架】

基本框架如图 4.1 所示，通过本项目的学习，能将图 4.1 的框架图转换成智慧工厂能源管理模块，效果图如图 4.2 所示。

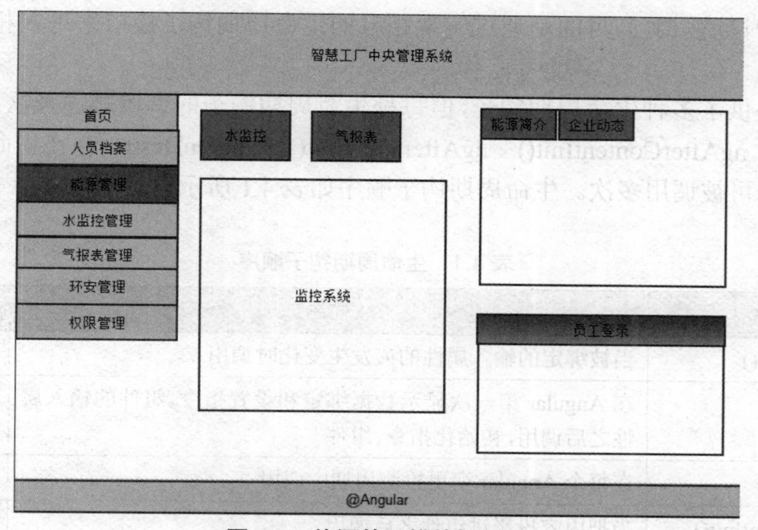

图 4.1　能源管理模块框架图

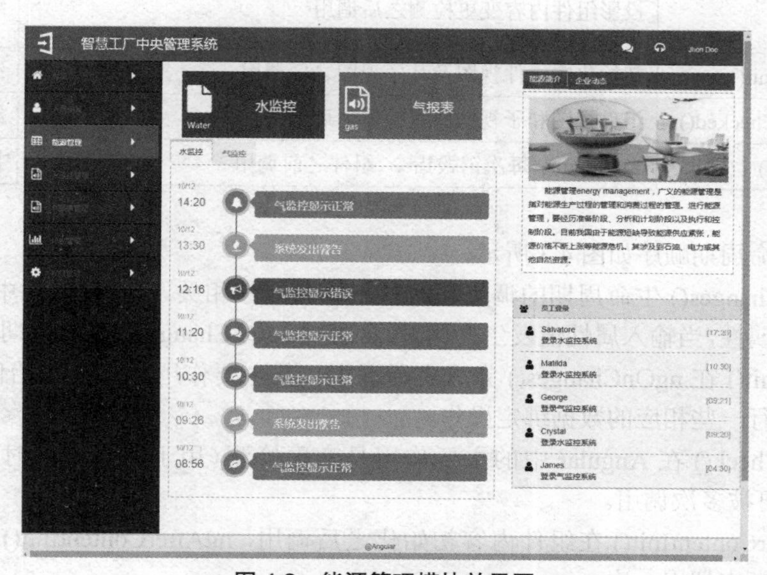

图 4.2　能源管理模块效果图

技能点 1 Angular 生命周期钩子

每个组件或指令都具备生命周期过程（新建→更新→销毁），并在生命周期的每个阶段之前或之后都可添加并执行相应的回调函数。对此，Angular 提供了生命周期钩子，在组件等生命周期的特定时刻被调用，目的是通过实现相应的生命周期钩子接口来进入相应的生命周期的关键时刻。

Angular 提供了多种生命周期钩子，但每种生命周期钩子的调用顺序及次数是不同的，其中 ngOnInit()、ngAfterContentInit()、ngAfterViewInit() 和 ngOnDestroy() 在生命周期中只会被调用一次，其他可被调用多次。生命周期钩子顺序如表 4.1 所示。

表 4.1 生命周期钩子顺序

名称	描述	范围
ngOnChanges()	当被绑定的输入属性的值发生变化时调用	指令和组件
ngOnInit()	在 Angular 第一次显示数据绑定和设置指令、组件的输入属性之后调用，初始化指令、组件	指令和组件
ngDoCheck()	在每个 Angular 变更检测周期中调用	指令和组件
ngAfterContentInit()	当把内容投影进组件之后调用	组件
ngAfterContent-Checked()	投影组件内容变更检测之后调用	组件
ngAfterViewInit()	初始化完组件视图及其子视图之后调用	组件
ngAfterViewChecked()	组件视图和子视图变更检测之后调用	组件
ngOnDestroy()	当 Angular 每次销毁指令、组件之前调用	指令和组件

组件的生命周期顺序如图 4.3 所示。

● ngOnChanges() 生命周期的调用与组件中输入属性相关。如：在组件中通过 @Input() 定义一个输入属性，当输入属性值发生变化时，会触发 ngOnChanges() 生命周期钩子。

● ngOnInit() 在 ngOnChanges() 完成之后调用。其主要作用是组件初始化时被调用，在调用期间可进行一些相应的数据绑定操作。

● ngDoCheck() 在 Angular 中的主要作用是变更检测（用于监听不同时期数据变化）。ngDoCheck() 可被多次调用。

● ngAfterContentInit() 在组件内容初始化之后调用，ngAfterContentInit() 生命周期钩子在整个生命周期中调用一次。

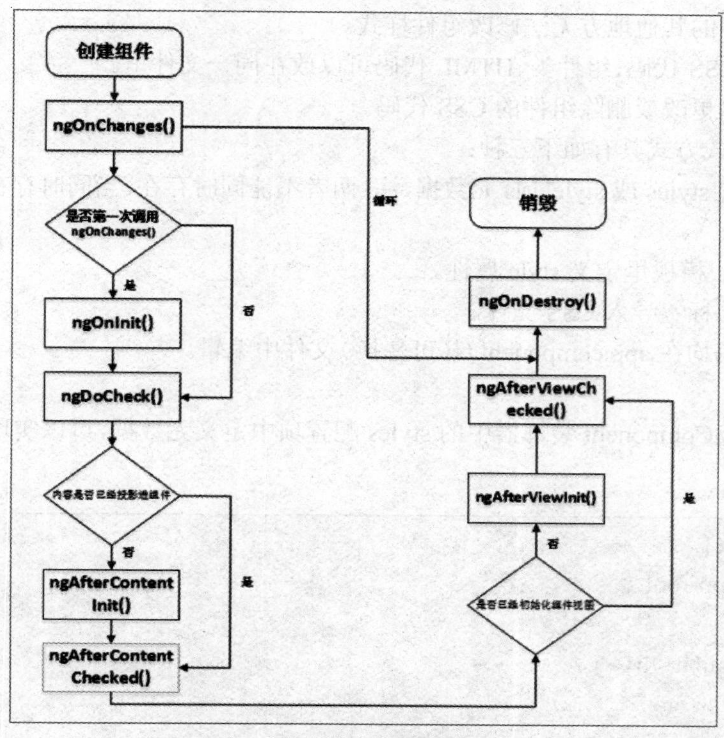

图 4.3 组件生命周期顺序图

- ngAfterContentChecked() 生命周期钩子在投影内容发生变化时被调用。与 ngDoCheck() 生命周期钩子类似，主要作用是变更检测。
- ngAfterViewInit() 当初始化完组件及其子组件并渲染呈现在页面上时，开始调用 ngAfterViewInit() 生命周期钩子。在整个组件生命周期中 ngAfterViewInit() 生命周期钩子只会调用一次。
- ngAfterViewChecked() 当父组件和子组件同时发生视图变化时调用，将先执行子组件的生命周期钩子。
- ngOnDestroy() 在生命周期结束进行组件或指令的销毁。

技能点 2 Angular 组件

1 Angular 组件样式

在 Angular 中，为了界面的美观，可以直接使用 CSS 对组件模板进行美化。在组件中，设置组件样式有以下优势：

- 支持 CSS 类名和选择器，且支持当前组件。
- CSS 类名和选择器不会与应用程序中其他类和选择器相冲突。

- 应用程序的其他地方无法修改组件样式。
- 组件的 CSS 代码、组件类、HTML 代码可以放在同一文件里。
- 随时可以更改或删除组件的 CSS 代码。

组件添加样式方式具有如下三种：

- 通过设置 styles 或 styleUrls 元数据，注：两者不能同时存在，当同时存在时，styles 中数据无作用。
- 在 HTML 模板里定义 style 属性。
- 通过 link 标签导入 CSS 文件。

注：以下代码均在 app.component.ts（可替换）文件中编辑。

（1）元数据

➢ 通过在 @Component 装饰器中的 styles 配置项中定义元数据，可以实现样式更改，代码如下所示。

```
@Component({
  selector: ' app-root ',
  template: `
    <h1>Angular</h1>
`,
  styles: ['h1 {
          font-weight: bold;
      }'],
})
```

➢ 通过在 @Component 装饰器中添加 styleUrls 配置项，可以引入 CSS 文件来修改样式，代码如下所示。

```
@Component({
selector: 'app-root',
template: `
<h2> Angular </h2>
`,
styleUrls: ['./app.component.css']
})
```

引入的 CSS 文件（app.component.css 文件）代码如下所示。

```
h1 {
  color:red;
  font-size:20px;
}
```

（2）模板内定义 style 标签

通过在组件的 HTML 模板中添加 <style> 标签定义样式，代码如下所示。

```
@Component({
  selector: ' app-root ',
  template: `
    <style>
      button {
        background-color: white;
        border: 1px solid #777;
      }
    </style>
    <h3>Angular</h3>
    <button>Activate</button>
    `
})
```

（3）模板内定义 link 标签

与 styleUrls 实现方式类似，都是引入外部 CSS 文件。不同的是，<link> 标签在 HTML 模板中使用，也可在 index.html 文件中使用，代码如下所示。

```
@Component({
  selector: ' app-root ',
  template: `
    <link rel="stylesheet" href="app.component.css">
    <h3>Team</h3>
    `
})
```

引入的 CSS 文件（app.component.css 文件）代码如下所示。

```
h3 {
  background-color:blue;
}
```

使用 Angular 组件样式效果如图 4.4 所示。

Angular组件样式（通过styleUrls导入CSS样式）

Angular组件样式（通过定义style标签设置）

图 4.4　组件样式

为了实现图 4.4 效果，代码如 CORE0401 所示。

代码 CORE0401：组件样式

```
import { Component } from '@angular/core';
@Component({
  selector: 'app-root',
  template: `
    <style>
     h1{
       background-color: yellow;
       border: 1px solid #777;
       font-size:20px
     }
    </style>
    <h2> Angular 组件样式（通过 styleUrls 导入 CSS 样式）</h2>
    <h1> Angular 组件样式（通过定义 style 标签设置）</h1>
  `,
  styleUrls: ['./project/project.component.css']
})
export class AppComponent {
}
```

导入的 CSS 文件代码如 CORE0402 所示。

代码 CORE0402：project.component.css

```
h2{color:red}
```

2 Angular 动态组件

为了吸引更多的用户，Angular 提供了动态组件，使项目在运行期间不仅可以引入固定组件，还可以动态加载组件，其具有使界面美观、灵活等优点。通过图 4.5、图 4.6 所示动态组件效果图学习如何创建动态组件。

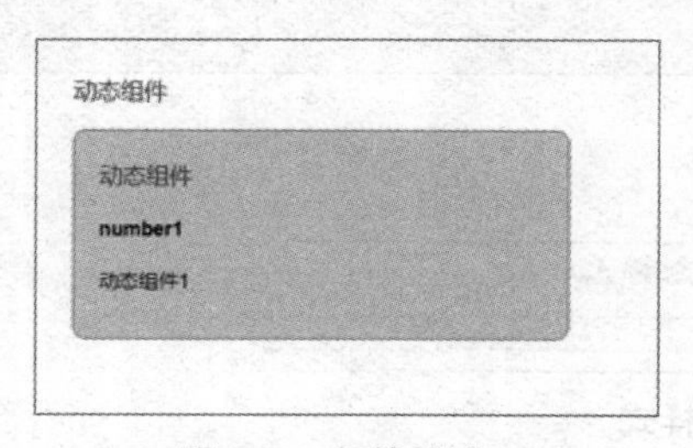

图 4.5 组件样式 1

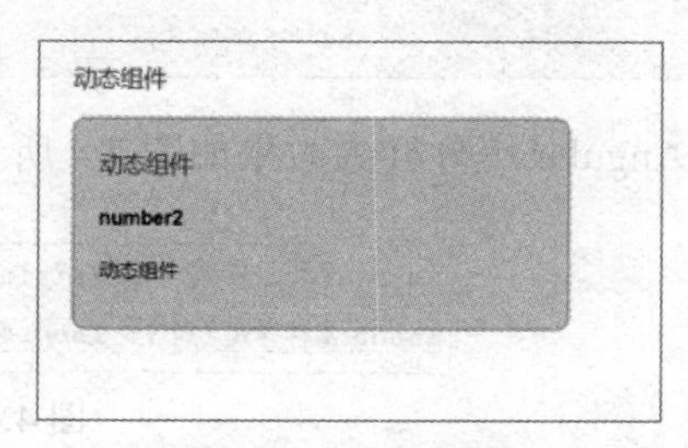

图 4.6 组件样式 2

第一步：指令标记组件显示位置

在创建动态组件之前，首先通过命令（ng g directive ad）创建指令文件，配置指令，通过类的指令在模板中标记出有效的插入点，以便获得组件的添加位置。代码如下所示。

```
import { Directive, ViewContainerRef } from '@angular/core';
@Directive({
    selector: '[ad-host]',
     })
     export class AdDirective {
     // ViewContainerRef 用于表示一个视图容器，可添加一个或多个视图，主要作用是
创建和管理内嵌视图或组件视图
     constructor(public viewContainerRef: ViewContainerRef) { }
}
```

第二步：加载组件

在组件中创建一个模板元素 <ng-template>，插入指令，代码如下所示。

```
import { Component, Input, AfterViewInit, ViewChild, ComponentFactoryResolver, OnDestroy }
 from '@angular/core';

import { AdDirective } from './ad.directive';
import { AdItem }      from './ad-item';
import { AdComponent } from './ad.component';

@Component({
  selector: 'app-add-banner',
  template: `
                <div class="ad-banner">
                <h3> 动态组件 </h3>
                <ng-template ad-host></ng-template>
                </div>
                `
})
```

第三步：解析组件

创建一个组件接收来自一个 AdService 的 AdItem 对象数组作为输入，AdBannerComponent 可以循环遍历 AdItems 的数组，通过 getAds() 方法，每三秒调用一次 loadComponent() 来加载新组件。

```
// AdBannerComponent 接收一个 AdItem 对象的数组作为输入
export class AdBannerComponent implements AfterViewInit, OnDestroy {
// AdItem 对象指定要加载的组件类，以及绑定到该组件上的任意数据
@Input() ads: AdItem[];
 currentAddIndex: number = -1;
@ViewChild(AdDirective) adHost: AdDirective;
subscription: any;
 interval: any;
constructor(private componentFactoryResolver: ComponentFactoryResolver) { }
ngAfterViewInit() {
this.loadComponent();
this.getAds();
 }
 ngOnDestroy() {
clearInterval(this.interval);
}
// loadComponent() 使用循环选取算法
 loadComponent() {
 this.currentAddIndex = (this.currentAddIndex + 1) % this.ads.length;
let adItem = this.ads[this.currentAddIndex];
// 使用 ComponentFactoryResolver 来为每个具体的组件解析出一个 Component-
Factory。 然后 ComponentFactory 会为每一个组件创建一个实例
 let componentFactory =
 this.componentFactoryResolver.resolveComponentFactory(adItem.component);
 let viewContainerRef = this.adHost.viewContainerRef; viewContainerRef.clear();
 let componentRef =
viewContainerRef.createComponent(componentFactory);
      (<AdComponent>componentRef.instance).data = adItem.data;
}

// 通过 getAds() 方法，每三秒调用一次 loadComponent() 来加载新组件
 getAds() {
this.interval = setInterval(() => { this.loadComponent(); }, 3000);
 }
}
```

第四步：创建公共接口

所有组件都实现了一个公共接口 AdComponent，它定义了一个标准化的 API，可以把数据传给组件。

```
export interface AdComponent {
    data: any;
}
```

第五步：组件调用接口

组件可通过以下代码调用公共接口，实现数据传输。代码如 CORE0403 所示。

代码 CORE0403：hero-profile.component.ts

```
import { Component, Input }  from '@angular/core';
import { AdComponent }       from './ad.component';
@Component({
  template: `
    <div class="hero-profile">
      <h3> 动态组件 </h3>
      <h4>{{data.name}}</h4>
      <p>{{data.bio}}</p>
    </div>
  `
})
export class HeroProfileComponent implements AdComponent {
  @Input() data: any;
}
```

快来扫一扫！

提示：当我们学会生命周期钩子基础知识后，是否会使用生命周期钩子呢？扫描图中二维码，使你对生命周期钩子拥有更全面的了解。

技能点 3　Angular 内置指令

指令是 Angular 中最常用的功能之一，使用指令可以最大程度的减少 DOM 操作，实现数据绑定和业务逻辑的交互。指令拓展并增强了 HTML 语法，可在 DOM 元素中添加或修改某些功能。Angular 提供了很多功能强大又实用的指令。其通用指令主要包括 NgClass、NgStyle、NgSwitch 等。

1 NgClass

NgClass 指令主要作用是可以动态设置和编辑给定元素的 CSS 类，通常在 HTML 模板中用 ngClass 表示。通过绑定 NgClass 指令，可以同时添加或移除多个类。NgClass 指令具有多种用法，部分如表 4.2 所示。

表 4.2 NgClass 指令用法

用法	描述
[ngClass]="first second"	给定一个元素添加类名，通过类名定义内容
[ngClass]="{'first':true,'second':true,'third': false}"	通过布尔值进行判断，显示哪种内容
[ngClass]="stringExp|arrayExp|objExp"	使用对象、字符串、数组来添加多个类
[ngClass]="{'class1 class2 class3' : true}"	可批量添加类名

使用 NgClass 指令如图 4.7 所示。

NgClass指令
NgClass指令
saveable | modified: | special: Refresh currentClasses
This div should be saveable, unchanged and, special after clicking "Refresh".
Ngclass指令
Ngclass指令
Ngclass指令

图 4.7 NgSwitch 指令

使用 NgClass 指令 HTML 代码如 CORE0404 所示。

代码 CORE0404：NgClass 指令

```
<h2 id="ngClass">NgClass 指令 </h2>
<div [ngClass]="currentClasses">NgClass 指令 </div>
<br>
<label>saveable <input type="checkbox" [(ngModel)]="canSave"></label> |
<label>modified: <input type="checkbox"
 [value]="!isUnchanged"(change)="isUnchanged=!isUnchanged"></label> |
<label>special:   <input type="checkbox" [(ngModel)]="isSpecial"></label>
<button (click)="setCurrentClasses()">Refresh currentClasses</button>
<div [ngClass]="currentClasses">
 This div should be {{ canSave ? "": "not"}} saveable,
                    {{ isUnchanged ? "unchanged" : "modified" }} and,
                    {{ isSpecial ? " ": "not"}} special after clicking "Refresh".</div>
```

```
<div [ngClass]="isSpecial ? 'special' : '' ">Ngclass 指令 </div>
<div class="bad curly special">Ngclass 指令 </div>
<div [ngClass]="{'bad':false, 'curly':true, 'special':true}">Ngclass 指令 </div>
```

对应的部分 ts 代码如 CORE0405 所示。

代码 CORE0405：ts 文件

```
ngOnInit() {
  this.setCurrentClasses();
}
currentClasses: {};
setCurrentClasses() {
// 动态控制类名
  this.currentClasses = {
    'saveable': this.canSave,
    'modified': !this.isUnchanged,
    'special': this.isSpecial
  };
}
```

2　NgStyle

NgStyle 指令的主要作用是可以使用动态值给特定的 DOM 元素设定 CSS 属性。在 Angular 中，使用 NgStyle 指令可同时设置多个内联样式。其实现过程需要 NgStyle 指令绑定到一个 key:value 控制对象（key 是样式名，value 是能用于这个样式的任何值）。NgStyle 指令使用方式具有多种，部分如表 4.3 所示。

表 4.3　NgStyle 指令用法

用法	描述
[style.background-color=" 'yellow' "]	设置元素的背景颜色
[ngStyle]="{color:'red'}" [style.font-size.px]="fontSize"	设置元素的字体大小、颜色
[ngStyle]="{color:'white', 'background-color': 'blue'}"	设置 DOM 元素的样式（通过键值对的形式）

技能点 4　Angular 自定义指令

当内置指令无法满足需要时可以创建自定义指令。自定义指令有自定义属性和自定义结构指令，其具有灵活、便捷等优点。本项目主要介绍自定义属性指令。

每个自定义属性指令都需要一个控制器类，在实现属性指令时，需通过 @Directive 装饰器（声明当前类是一个指令）将元数据添加到控制器类上，控制器类则实现了指令所对应的特定行为。

在元数据中声明 selector 属性用以标志指令的选择对象，selector 属性写法如表 4.5 所示。

表 4.5 selector 属性写法

用法	描述
element-name	按元素名称选择
.class	按类名称选择
[attribute]	按属性名称选择
[attribute=value]	使用属性和值选择
:not(sub_selector)	只有当元素与 sub_selector 不匹配时才选择
selector1,selector2	选择 selector1 或 selector2

使用自定义指令实现在输入框输入空格时，默认删除的效果，如图 4.8 所示，指令步骤如下所示。

图 4.8 自定义指令

第一步：创建自定义指令文件

在命令窗口输入以下命令，创建自定义指令文件。命令如下所示。

```
ng g directive project
```

第二步：创建自定义指令

在新建的自定义指令文件中定义 @Directive 装饰器，指定指令所关联的属性选择器。代码如 CORE0406 所示。

代码 CORE0406：project.directive.ts 文件

```
// 从 @angular/core 模块中引入 Directive 和 ElementRef
import {Directive}
 from "@angular/core";
import { ElementRef, HostListener} from "@angular/core";
@Directive({
```

```
// appMyDirective 为元素名称
   selector: '[appProject]'
})
// 该类实现了指令所包含的逻辑
export class ProjectDirective {
   //ElementRef 用来访问 DOM 元素
   constructor(public elementRef: ElementRef) {
      // 设置元素背景颜色
      elementRef.nativeElement.style.backgroundColor='lightblue';
   }
 // HostListener 是属性装饰器，用来监听 keyup 事件
// 当表单中有输入时会调用方法，传递 $event 对象进去
 @HostListener('keyup', ['$event.target'])
 // 判断是否有值输入
 keyupFun(evt) {
    if (evt.value) {
        //ElementRef 需要在构造函数中注入进去 ,trim() 方法去除空格
        this.elementRef.nativeElement.value = evt.value.trim();
    }
   }
}
```

第三步：在模块中声明

指令创建成功后需在模块中声明才可使用，代码如 CORE0407 所示。

代码 CORE0407：app.module.ts 文件

```
import { BrowserModule } from '@angular/platform-browser';
import { NgModule } from '@angular/core';
import { AppComponent } from './app.component';
import { ProjectDirective } from './project.directive';
@NgModule({
  declarations: [
    AppComponent,
    ProjectDirective
  ],
  imports: [
    BrowserModule
  ],
```

```
  providers: [],
  bootstrap: [AppComponent]
})
export class AppModule { }
```

第四步：在模板文件中引用指令

在该组件的 HTML 模板元素中添加对应的属性，代码如 CORE0408 所示。

代码 CORE0408：app.component.html

```
// appProject 绑定输入变量的属性
<h1 appProject>Angular 自定义指令 </h1>
<label for="name"> 姓名 </label>
<input type="text" id="name" appProject>
```

快来扫一扫！

提示：Angular 应用是模块化的，并且 Angular 有自己的模块系统，被称为 Angular 模块或 NgModule。扫描图中二维码，了解更多 Angular 模块相关信息。

通过下面七个步骤的操作，实现图 4.2 所示的智慧工厂能源管理模块的效果。

第一步：将能源管理模块分为能源简介部分、导航部分、员工登录部分、监控系统部分。

第二步：创建能源模块，主组件和能源简介、导航、员工登录、监控系统等子组件。

第三步：在主组件 energy-management.component.html 中对模块进行布局，设置各个组件的渲染位置。代码如 CORE0409 所示。

代码 CORE0409：页面布局

```
<div style="height:100%;padding-bottom: 30px;background-color: white;">
<div class="row">
```

```
  <div class="col-md-7">
    <!-- 水监控、气报表导航 -->
    <app-enery-carousel></app-enery-carousel>
    <!-- 监控系统 -->
    <app-enery-project></app-enery-project>
  </div>
  <div class="col-md-5">
    <!-- 能源简介 -->
    <app-enery-select></app-enery-select>
    <!-- 员工登录 -->
    <app-enery-user></app-enery-user>
 </div>
</div>
</div>
```

第四步：路由配置，在 energy-management.module.ts 模块文件中配置路由，代码如 CORE0410 所示。

代码 CORE0410：路由配置

```
import {GasReportingComponent}
from "../gas-reporting/gas-reporting/gas-reporting.component";
import {WaterMonitoringComponent}
from "../water-monitoring/water-monitoring.component";
import { RouterModule, Routes } from '@angular/router';
//...
const appRoutes: Routes = [
  {path:'waterMonitoring',component:WaterMonitoringComponent},
  {path:'gasReporting',component:GasReportingComponent},
];
@NgModule({
  imports: [
    //...
    RouterModule.forRoot(appRoutes),
  ],
  //...
export class EnergyManagementModule { }
```

在导航组件中，使用路由，通过点击，跳转到相应的页面中，代码如 CORE0411 所示。设置样式前效果如图 4.9 所示。

代码 CORE0411：路由跳转

```
<div class="metro-nav metro-fix-view" style="margin-top: 20px;margin-left: 20px">
<div class="metro-nav-block nav-block-blue double" style="width: 50%" >
  <a  routerLink="/waterMonitoring">
    <i class="fa fa-file"></i>
    <div class="info"> 水监控 </div>
    <div class="status">Water</div>
  </a>
</div>
  <div class="metro-nav-block nav-block-red double" style="width: 50%;margin-left: 20px" >
    <a  routerLink="/gasReporting">
      <i class="fa fa-file-audio-o"></i>
      <div class="info"> 气报表 </div>
      <div class="status">gas</div>
     </a>
  </div>
</div>
```

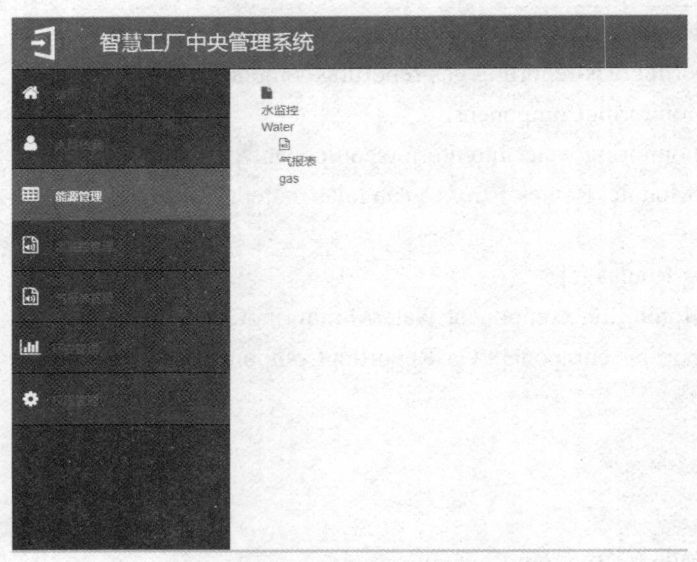

图 4.9　设置样式前

通过 CSS 样式设置导航的动态效果，为其添加背景颜色并设置字体大小等，代码如 CORE0412 所示。设置样式后效果如图 4.10 所示。

代码 CORE0412：CSS 样式

```
/* 字体样式 */
.metro-nav {
    font-family: 'MyriadPro-Light';
    position: relative;
    z-index: 101;
}
.metro-fix-view .metro-nav-block.double {
    width: 251px !important;
}
.metro-fix-view .metro-nav-block {
    width: 124px !important;
}
/* 显示位置 */
.metro-fix-view .metro-nav-block {
    float: left;
    height: 120px;
    margin: 0 1% 1% 0;
    position: relative;
}
/* 背景颜色 */
.metro-nav .nav-block-blue {
    background: #2e6da4;
}
.metro-nav .nav-block-red {
    background:#4cae4c;
}
.metro-fix-view .metro-nav-block a {
    width: 84% !important;
}
.metro-nav .metro-nav-block a {
    color: white;
    height: 90%;
    padding: 5px 10px;
    position: absolute;
}
.metro-nav .metro-nav-block a i {
    transition: all 0.5s ease-in-out 0s;
```

```
}
.metro-nav .metro-nav-block i {
   font-size: 50px;
   margin-top: 20px;
   display: inline-block;
}
/* 动画过渡 */
.metro-nav .metro-nav-block a .info {
   ransition: all 0.4s ease-in-out 0s;
}
.metro-nav .metro-nav-block .info {
   font-size: 24px;
   position: absolute;
   right: 10px;
   top: 45px;
}
.metro-nav .metro-nav-block .status, .metro-nav .metro-nav-block .tile-status {
   background-color: transparent;
   bottom: -10px;
   font-size: 14px;
   left: 10px;
   min-height: 30px;
   position: absolute;
}
/* 文字动画效果 */
.metro-nav .metro-nav-block a:hover .info {
   transform:rotate(360deg);
   font-size: 40px;
   opacity: 1;
}
/* 字体图标动画效果 */
.metro-nav .metro-nav-block a:hover i{
   transform:rotate(83deg);
   font-size: 140px;
   opacity: 0;
}
```

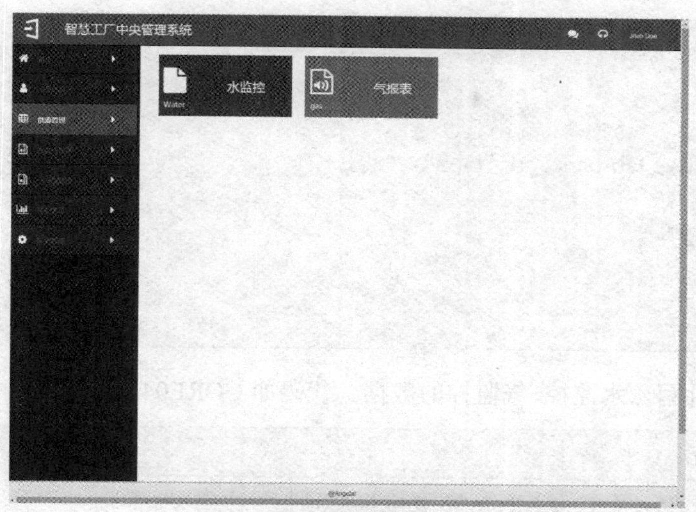

图 4.10　设置样式后

第五步：监控系统分为气、水监控，通过选项卡形式分别对两个系统进行监控。监控的数据通过 NgFor 指令双向绑定显示。部分代码如 CORE0413 所示。设置样式前效果如图 4.11 所示。

代码 CORE0413：监控系统

```
<div style="padding-top: 50px;">
<ul class="nav nav-tabs" >
   <li class="active"><a href="#tab1" data-toggle="tab"> 水监控 </a></li>
   <li ><a href="#tab2" data-toggle="tab"> 气监控 </a></li>
</ul>
<div class="tab-content" >
   <div class="tab-pane" id="tab1">
      <div class="row-fluid">
         <ul class="metro_tmtimeline"  *ngFor="let product of products1">
            <li class="{{product.color}}">
               <div class="metro_tmtime" >
                  <span class="date">{{product.date}}</span>
                  <span class="time" >{{product.time}}</span>
               </div>
               <div class="metro_tmicon">
                  <i class="fa {{product.iconclass}}"></i>
               </div>
               <div class="metro_tmlabel" style="height: 10px;padding-top: 15px;">
                  <h2 >{{product.work}}</h2>
               </div>
            </li>
```

```
        </ul>
    </div>
</div>
  <div class="active tab-pane" id="tab2">
  //…
  </div>
</div>
</div>
```

在对应 ts 文件写入水监控、气监控的数据。代码如 CORE0414 所示。

代码 CORE0414：ts 代码

```
import { Component, OnInit } from '@angular/core';
@Component({
  selector: 'app-enery-project',
  templateUrl: './enery-project.component.html',
  styleUrls: ['./enery-project.component.css']
})
export class EneryProjectComponent implements OnInit {
  private  products1: Array<Product>;
  private  products2: Array<Product>;
  constructor() { }
  ngOnInit() {
        this.products1=[
          new Product("10/12","14:20"," 水监控显示正常 ","Salvatore","XX 电监控成
                功 ",1,"blue","fa-bell"),
          new Product("10/12","13:30"," 水监控显示正常 ","Matilda","XX 电监控成
                功 ",2,"blue","fa-fire"),
// 部分代码省略
        ]
      }
}
// 定义一个类
export class Product{
  constructor(
      public date:string,
      public time:string,
      public work:string,
```

```
        public name:string,
        public success:string,
        public id:number,
        public color:string,
        public iconclass:string
    )
    {
    }
}
```

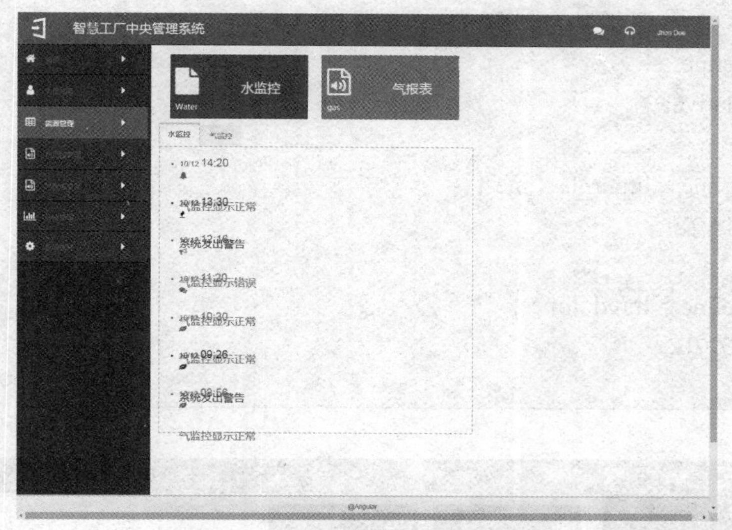

图 4.11　设置样式前

设置监控系统样式，通过为监控时间设置字体颜色，设置监控的图标等样式。部分代码 CORE0415 如下，设置样式后效果如图 4.12 所示。

代码 CORE0415：CSS 代码

```
.metro_tmtimeline {
    margin: 30px 0 0 0;
    padding: 0;
    list-style: none;
    position: relative;
}
.metro_tmtimeline:before {
    content: ' ';
    position: absolute;
```

```
    top: 0;
    bottom: 0;
    width: 2px;
    background: dodgerblue;
    left: 20%;
    margin-left: -6px;
    height: 100px;
}
.metro_tmtimeline:last-child:before {
    height: 0px;
}
.metro_tmtimeline > li {
    position: relative;
}
.metro_tmtimeline > li.purple .time {
    color: #9D4A9C;
}
.metro_tmtimeline > li.red .time {
    color: #DE577B;
}
```

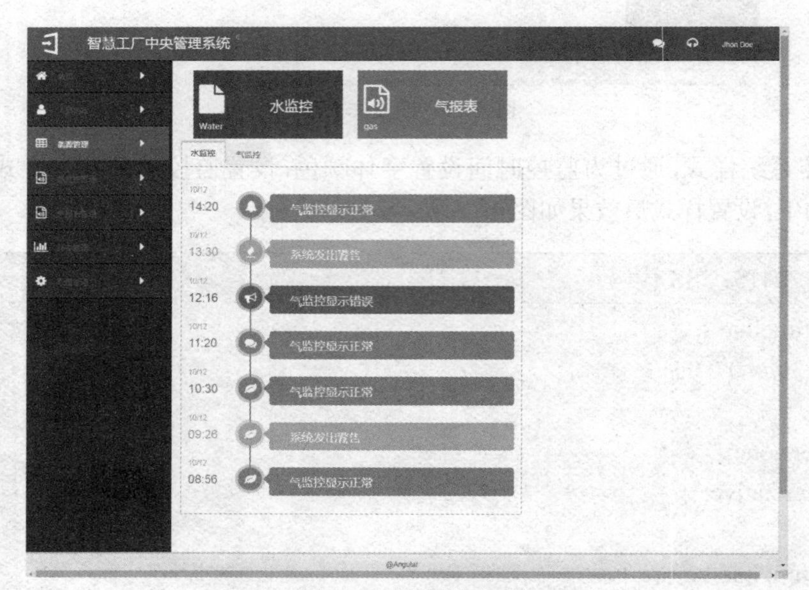

图 4.12　设置样式后

第六步：能源简介组件。使用 ul 列表设置导航，通过点击 li 元素进行切换。部分代码如 CORE0416 所示。设置样式前效果如图 4.13 所示。

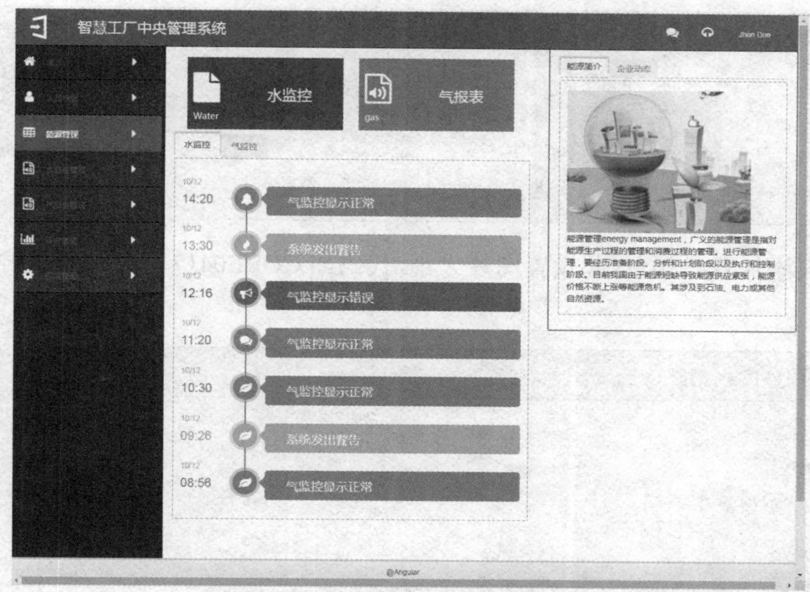

图 4.13　设置样式前

```
代码 CORE0416：能源简介
<div class="row-fluid " >
  <div >
    <div class="widget "  >
    <div class="widget-body" >
    <ul class="nav nav-tabs"  id="myTab" >
      <li class="active"  ><a href="#tab_1_1" data-toggle="tab"  > 能源简介 </a></li>
      <li ><a href="#tab_1_2" data-toggle="tab" > 企业动态 </a></li>
    </ul>
    <div class="tab-content" >
      <div class="tab-pane active" id="tab_1_1">
        <div class="tab-img" >
          <img src="../../../assets/0.jpg" >
            </div>
            <div class="tab-text" > 能源管理 energy management，广义的能源管理是指对能源生产过程的管理和消费过程的管理。进行能源管理，要经历准备阶段、分析和计划阶段以及执行和控制阶段。目前我国由于能源短缺导致能源供应紧张，能源价格不断上涨等能源危机。其涉及到石油、电力或其他自然资源。</div>
            </div>
            <div class="tab-pane" id="tab_1_2">
            //…
```

```
            </div>
        </div>
    </div>
</div>
</div>
```

为 ul 列表添加背景颜色，并设置图片大小，代码如 CORE0417 所示。设置样式后效果如图 4.14 所示。

代码 CORE0417：CSS 样式

```
.widget{
  border:0px
 }
.widget-body{
  border:0px
}
#myTab{
  background-color: #3983c2
}
#myTab a{
  background-color: #3983c2;
  color: white
}
.tab-content{
  padding: 10px
}
.tab-content .tab-img{
  text-align: center
}
.tab-content .tab-img img{
  width: 350px;
  height: 150px;
}
.tab-content .tab-text{
  text-indent: 2em;
  line-height:25px;
  padding-top: 5px
}
```

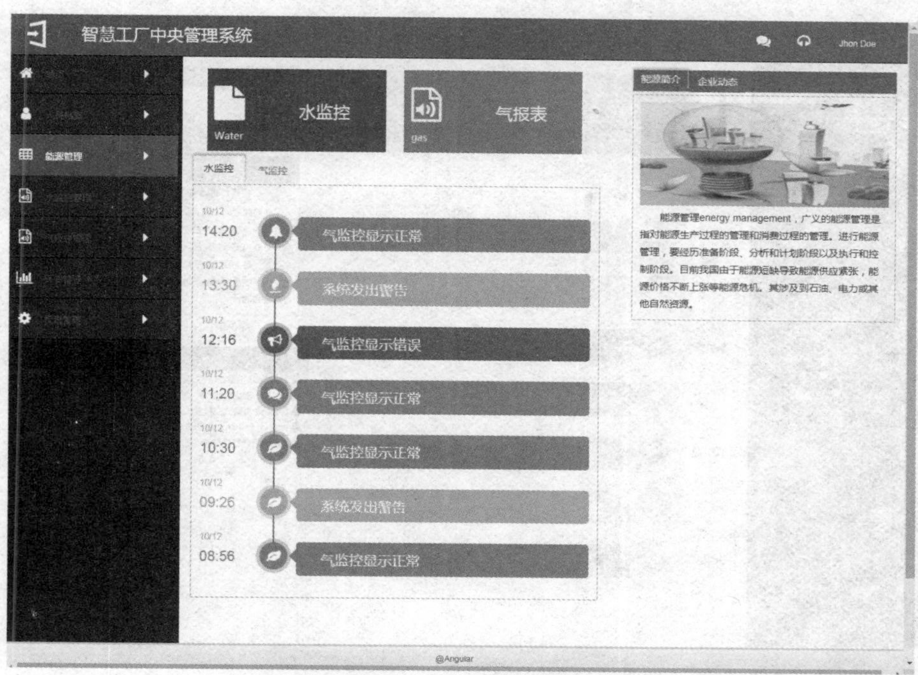

图 4.14 设置样式后

第七步：设置员工登录，通过员工登录模块可以查看到员工哪一时刻登录并查看监控系统。代码如 CORE0418 所示。设置样式前效果如图 4.15 所示。

代码 CORE0418：员工登录

```
<div class="widget-box" >
  <div class="widget-title bg_lo"  data-toggle="collapse" href="#collapseG3" >
  <span class="icon"> <i class="icon-chevron-down"></i> </span>
    <h5> 员工登录 </h5>
  </div>
  <div class="widget-content nopadding updates collapse in" id="collapseG3"
       style="padding-left: 10px;padding-right:20px "  *ngFor="let product of products">
    <div class="new-update clearfix" style="border-bottom: solid 1px #d2d2d2">
      <i class="icon-user"></i>
    <span class="update-notice">
      <span>{{product.name}}</span>
      <span>{{product.work}}</span>
    </span>
    <span class="update-date">[{{product.time}}]</span>
  </div>
  </div>
</div>
```

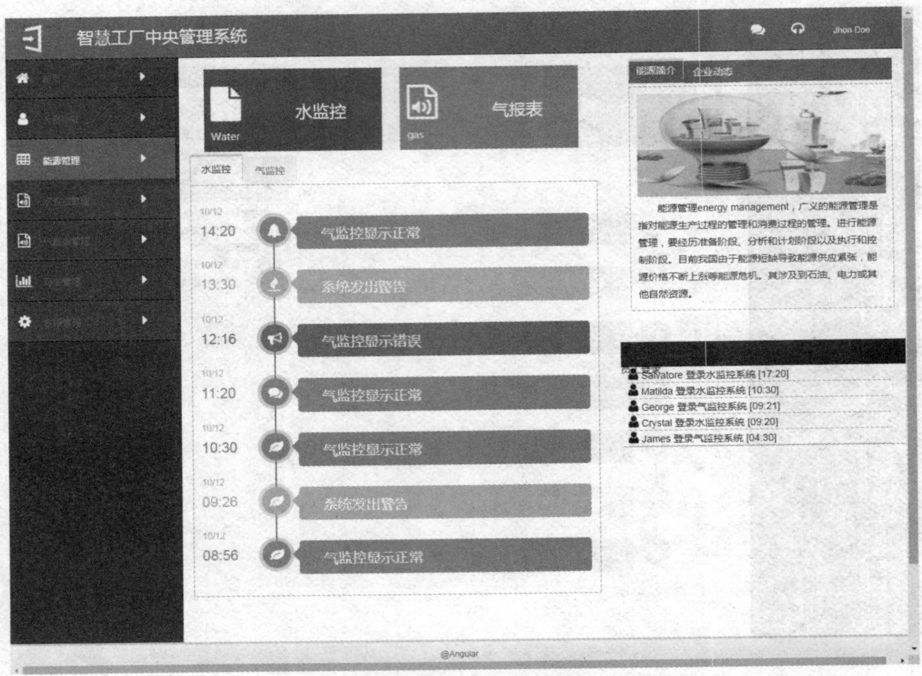

图 4.15　设置样式前

在对应 ts 文件写入数据并绑定在 HTML 模板中。代码如 CORE0419 所示。

代码 CORE0419：ts 代码

```
import { Component, OnInit } from '@angular/core';
@Component({
  selector: 'app-enery-user',
  templateUrl: './enery-user.component.html',
  styleUrls: ['./enery-user.component.css']
})
export class EneryUserComponent implements OnInit {
  private  products1: Array<Product1>;
  constructor() { }
  ngOnInit() {
    this.products1=[
      new Product1("10/12","17:20"," 登录水监控系统 ","Salvatore","XX 电监控成
          功 ",1,"green","fa-bell"),
      new Product1("10/12","10:30"," 登录水监控系统 ","Matilda","XX 电监控成
          功 ",2,"purple","fa-fire"),
      new Product1("10/12","09:21"," 登录气监控系统 ","George","XX 电监控成
          功 ",3,"red","fa-bullhorn"),
```

```
        new Product1("10/12","09:20"," 登录水监控系统 ","Crystal","XX 电监控成
            功 ",1,"yellow","fa-comments-alt"),
        new Product1("10/12","04:30"," 登录气监控系统 ","James","XX 电监控成
            功 ",1,"blue","fa-leaf")
        ]
    }
}
// 定义一个类
export class Product1{
  constructor(
    public date:string,
    public time:string,
    public work:string,
    public name:string,
    public success:string,
    public id:number,
    public color:string,
    public iconclass:string
  )
  {
  }
}
```

设置员工登录样式，设置字体间距、颜色等。代码如 CORE0420 所示。设置样式后效果如图 4.16 所示。

代码 CORE0420：CSS 代码

```
.widget-box {
  background: none repeat scroll 0 0 #F9F9F9;
  border-left: 1px solid #CDCDCD;
  border-top: 1px solid #CDCDCD;
  border-right: 1px solid #CDCDCD;
  clear: both;
  margin-top: 16px;
  margin-bottom: 16px;
  position: relative;
}
.widget-box.widget-calendar, .widget-box.widget-chat {
```

```
    overflow:hidden !important;
}
.accordion .widget-box {
    margin-top: -2px;
    margin-bottom: 0;
    border-radius: 0;
}
.widget-box.widget-plain {
    background: transparent;
    border: none;
    margin-top: 0;
    margin-bottom: 0;
}
.widget-title, .modal-header, .table th, div.dataTables_wrapper .ui-widget-header {
    background:#efefef;
    border-bottom: 1px solid #CDCDCD;
    height: 36px;
}
.widget-title .nav-tabs {
    border-bottom: 0 none;
}
.widget-title .nav-tabs li a {
    border-bottom: medium none !important;
    border-left: 1px solid #DDDDDD;
    border-radius: 0 0 0 0;
    border-right: 1px solid #DDDDDD;
    border-top: medium none;
    color: #999999;
    margin: 0;
    outline: medium none;
    padding: 9px 10px 8px;
    font-weight: bold;
    text-shadow: 0 1px 0 #FFFFFF;
}
.widget-title .nav-tabs li:first-child a {
    border-left: medium none !important;
}
.widget-title .nav-tabs li a:hover {
```

```
    background-color: transparent !important;
    border-color: #D6D6D6;
    border-width: 0 1px;
    color: #2b2b2b;
}
.widget-title .nav-tabs li.active a {
    background-color: #F9F9F9 !important;
    color: #444444;
}
.widget-title span.icon {
    padding: 9px 10px 7px 11px;
    float: left; border-right:1px solid #dadada;
}
.widget-title h5 {
    color: #666;
    float: left;
    font-size: 12px;
    font-weight: bold;
    padding: 12px;
    line-height: 12px;
    margin: 0;
}
.new-update {
    border-top: 1px solid #DDDDDD;
    padding: 10px 12px;
}
.new-update:first-child {
    border-top: medium none;
}
.new-update span {
    display:block;
}
.new-update i {
    float: left;
    margin-top: 3px;
    margin-right: 13px;
}
.new-update .update-date {
```

```
    color: #BBBBBB;
    float: right;
    margin: 4px -2px 0 0;
    text-align: center;
    width: 30px;
}
.new-update .update-date .update-day {
    display: block;
    font-size: 20px;
    font-weight: bold;
    margin-bottom: -4px;
}
```

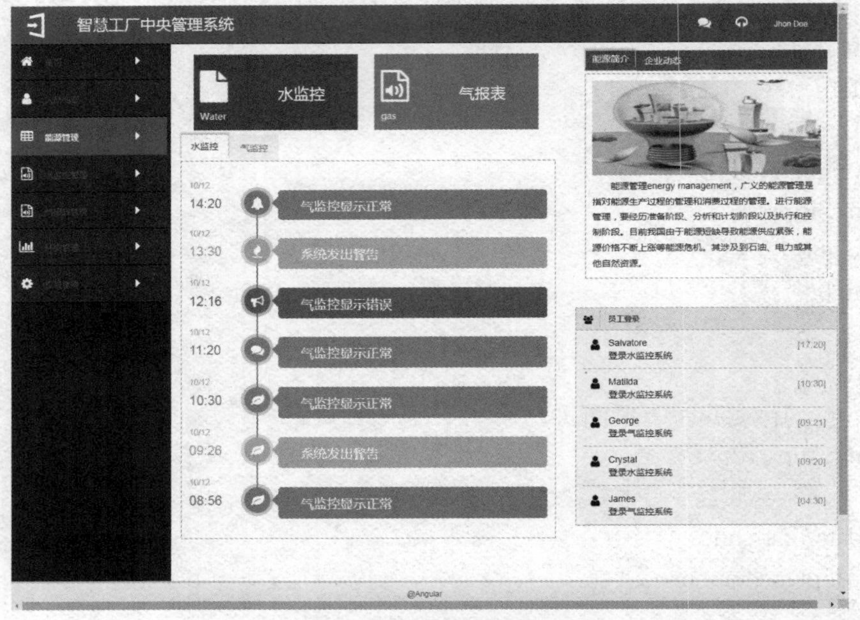

图 4.16 设置样式后

至此，智慧工厂能源管理模块制作完成。

本项目通过对智慧工厂能源管理模块的学习，对能源管理模块中各个组件的创建及所需功能具有进一步了解，掌握组件与指令生命周期钩子的使用方法及顺序，具有独立创建动态组件的能力，了解如何运用指令扩展浏览器的功能，为以后实现各个功能打下基础。

hook	生命周期
oninit	初始化
destroy	销毁
host	主机
component	组件
directive	指令
template	模块
switch	开关
selector	选择器

一、选择题

1. 下面对生命周期钩子描述错误的是（　）。

（A）ngAfterContentChecked() 生命周期钩子在投影内容发生变化时被调用

（B）当初始化完组件及其子组件时，开始调用 ngAfterViewInit() 生命周期钩子

（C）ngAfterViewChecked() 生命周期钩子在视图发生变化，执行完更检查机制后调用

（D）ngOnDestroy 生命周期钩子主要作用是反订阅可观察对象和分离事件处理器，以防内存泄漏

2. 下面生面周期钩子顺序正确的是（　）。

（A）ngAfterContentInit、ngAfterContentChecked、ngAfterViewInit、NgAfterViewChecked

（B）ngAfterContentChecked、ngAfterContentInit、ngAfterContentChecked、ngAfterViewInit

（C）ngAfterViewInit、ngAfterContentChecked、ngAfterContentInit、ngAfterContentChecked

（D）ngAfterContentChecked、ngAfterViewInit、ngAfterContentChecked、ngAfterContentInit

3. 下面对 NgStyle 用法错误的是（　）。

（A）[style.background-color=" 'yellow' "]

（B）[ngStyle]= "{color: 'red'}" [style.font-size.px]= "fontSize"

（C）[ngStyle]= "{color: 'white', 'background-color': 'blue'}"

（D）[ngStyle]= "{ 'white', 'blue'}"

4. 下面对组件样式说法错误的是（　）。

（A）支持 CSS 类名和选择器，且限当前组件上下文有意义

（B）CSS 类名和选择器不会与应用程序中其他类和选择器相冲突

（C）应用程序的其他地方可以修改组件样式

（D）组件的 CSS 代码、组件类、HTML 代码可以放在同一文件里

5. 下面对 NgClass 用法错误的是（　）。

（A）[ngClass]= " 'first second' "

（B）[ngClass]= " ['first', 'second'] "

（C）[ngClass]= "{'first': true, 'second': true, 'third': false}"

（D）[ngClass]= "{ true, true, false}"

二、填空题

1. 生命周期中的钩子调用先后顺序为 _____、_____、_____、_____、_____、_____、_____、_____。

2. ngOnChanges() 在 Angular 设置数据绑定输入属性时响应，且其生命周期的调用与 _____ 相关。

3.NgClass 指令通过在 Angular 的 HTML 模板中用 _______ 表示。

4.NgSwitch 实际上包括三个相互协作的指令：_______、_______ 和 _______。

5. 在实现属性指令时，需通过 ________ 将元数据添加到控制器类上。

三、上机题

使用 Angular 设置 NgSwitch 指令。

要求：利用 Angular 的 NgSwitch 指令，实现如图所示效果。

项目五　智慧工厂水监控模块

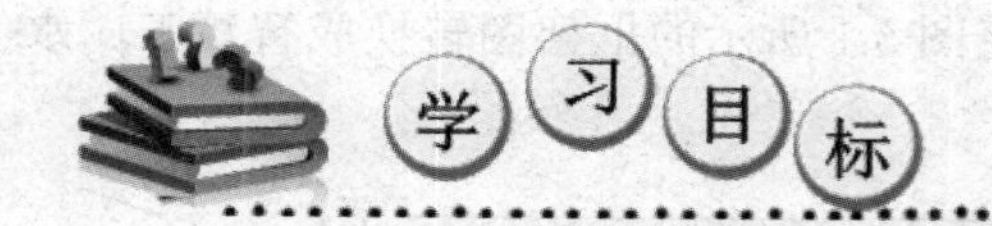

通过智慧工厂水监控模块功能的实现，了解如何注入服务，学习提供器与注入器的相关知识，掌握依赖注入、表单的应用，具有为应用注入服务的能力。在任务实现过程中：

- 了解依赖注入概念。
- 学习提供器与注入器的相关知识。
- 掌握依赖注入、表单的应用。
- 具有使用依赖注入为应用注入服务的能力。

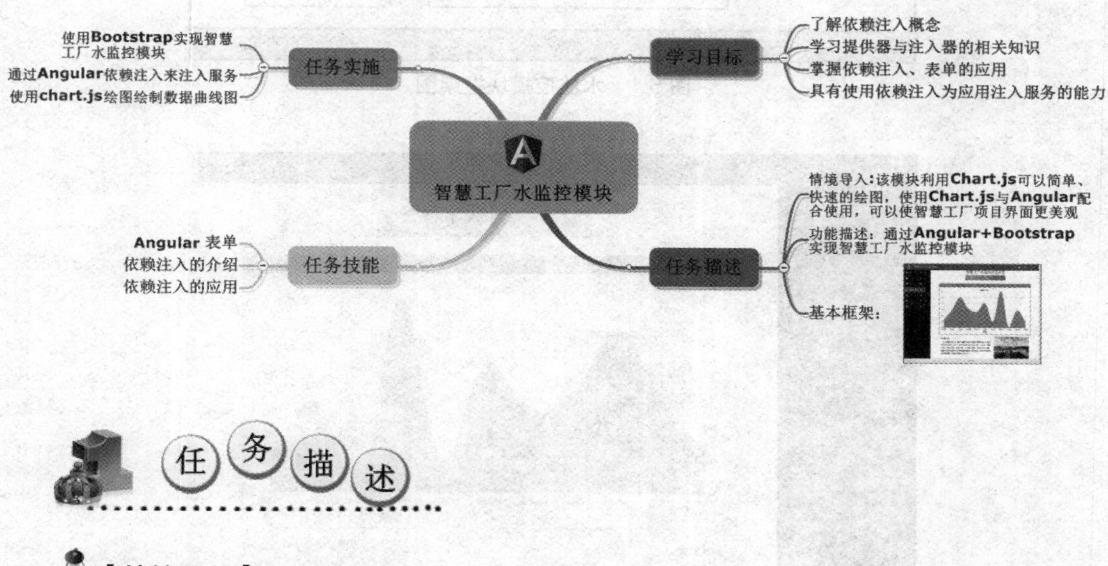

任务描述

【情境导入】

数据的统计方式具有很多种，为了更加直观的显示数据，该公司的开发人员选择了曲线统计图来实时监控水资源数据，工作人员可以清晰看出各个时间段水压数据的多少，这样节省了大量的查询时间，减少工作人员的工作量。本项目主要是通过实现智慧工厂的水监控模块来学习 Angular 的依赖注入。

【功能描述】

使用 Bootstrap+Angular 实现智慧工厂水监控模块：

- 使用 Bootstrap 实现智慧工厂水监控模块。
- 通过 Angular 依赖注入来注入服务。
- 使用 Chart.js 绘制数据曲线。

【基本框架】

基本框架如图 5.1 所示，通过本项目的学习，能将图 5.1 所示的框架图转换成智慧工厂水监控模块，效果图如图 5.2 所示。

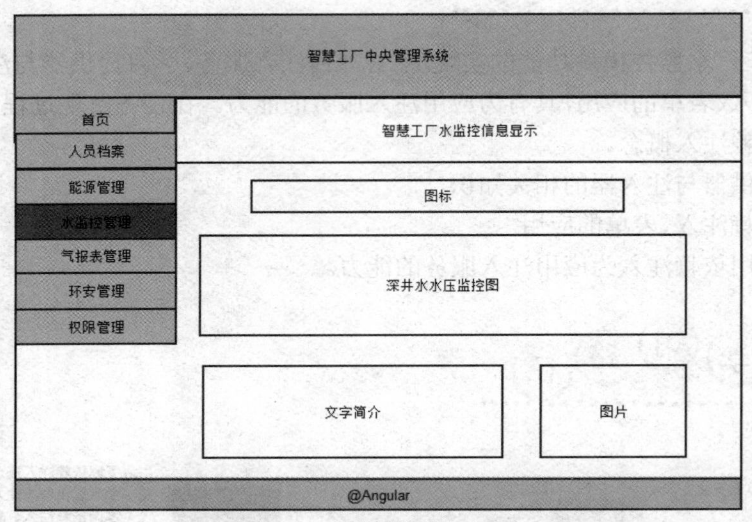

图 5.1　水监控模块框架图

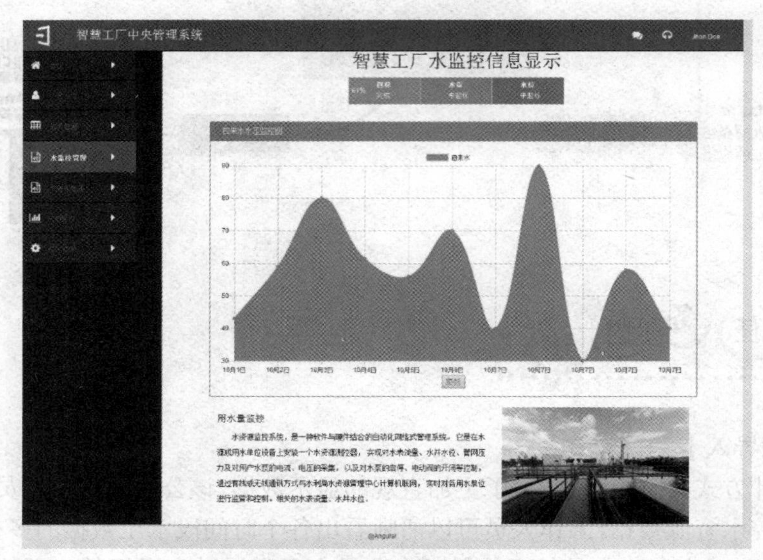

图 5.2　水监控模块效果图

技能点 1　Angular 表单

表单的使用场景十分广泛，常见场景有用户登录、注册等。虽然在 HTML 中内置了表单标签（input、select、textarea），但这些标签特性存在浏览器兼容问题，且自定义验证规则、表单数据处理等操作复杂。针对以上问题，Angular 团队对表单进行了封装扩展，提供了模板驱动、响应式两种方式构建表单。使开发者可以使用简洁的代码来构建功能强大的表单。

模板驱动表单的大部分相关代码都在模板里，通过在模板里面添加指令来定义模板和验证信息。本项目主要介绍模板驱动式表单。

1　表单中的指令

表单是 Web 程序中的重要组成部分，构建良好及实用的表单必不可少的是表单指令，表单指令是 Angular 对常用的表单交互功能进行封装扩展。在表单中常用的指令有 NgForm、NgModel、NgModelGroup 等，其使用之前都需在根模块下导入 FormModule 模块。

将 FormModule 模块添加到 @NgModule 元数据的 imports 数组中，使用 import 导入 FormModule，从而在整个应用的表单中都可以使用特有的表单指令。

```
import { BrowserModule } from '@angular/platform-browser';
import { NgModule } from '@angular/core';
import { AppComponent } from './app.component';
import { FormsModule } from "@angular/forms";
@NgModule({
  declarations: [
    AppComponent,
  ],
  imports: [
    BrowserModule,
    FormsModule
  ],
  providers: [],
  bootstrap: [AppComponent]
})
export class AppModule { }
```

（1）NgForm

NgForm 指令是表单的控制中心，负责处理表单内的业务逻辑，所有的表单指令都需要在 NgForm 指令内才能运行。在创建表单时，会在 <form> 标签上通过 # 变量名 ="ngForm" 添加 NgForm 指令，然后通过 (ngSubmit)="xxx(#ngForm.value)" 来获取表单的值。NgForm 指令为 form 增补了一些额外特性。它控制了标识 ngModel 指令和 name 属性的元素，并可以监听它们的属性变化（包括其有效性 valid）。

在 HTML 模板中定义 NgForm 指令。在表单中声明一个模板变量，在 <form> 标签中加入 #f="ngForm"（#f 变量名可变），代码如下所示。

```
<form #f="ngForm" (ngSubmit)="onSubmit(f.value) " >
        <div> 用户名：<input type="text" ></div>
        <div> 密码：<input type="password" ></div>
        <div> 确认密码：<input type="password" ></div>
        <button> 提交 </button>
</form>
```

注：#f="ngForm" 表示使用了一个名为 f 的变量引用了 NgForm 表单指令。

（2）NgModel

NgModel 指令是实现表单控件数据绑定的核心，表单添加了该指令后，Angular 会隐式创建一个 FormControl 类（表单控件容器），用于存储其值。但使用 NgModel 指令时，则必须设置 name 属性，其作为唯一标识符来生成一个 FormControl 类。

NgModel 指令支持单向和双向数据绑定，表单的单向数据绑定使用了 [ngModel]，双向数据绑定使用 [(ngModel)]。使用 NgModel 指令双向数据绑定代码如下所示。

```
@Component({
  selector: 'exe-app',
  template: `
  用户名：<input type="text" [(ngModel)]="username" name="username">
  {{username}}
 `,
})
export class AppComponent implements OnInit {
  username = 'Tom';
}
```

（3）NgModelGroup

NgModelGroup 指令是多个 ngModel 的集合，用于区分不同类型的表单控件。使其形成更清晰的层次关系。使用 NgModelGroup 指令代码如下所示。

```
<form #f="ngForm" (ngSubmit)="onSubmit(f.value)>
    <fieldset ngModelGroup="login">
      <legend> 登录 :</legend>
      Username: <input type="text" name="username" required>
      Password: <input type="password" name="password" requried>
    </fieldset>
    <fieldset ngModelGroup="address ">
      <legend> 省 :</legend>
      <select>
          <option> 河北 </option>
          <option> 河南 </option>
          <option> 北京 </option>
      </select>
  </fieldset>
</form>
```

2　表单样式

NgModel 指令不仅能追踪表单控件的状态，还可使用对应的 CSS 状态类来表示表单控件的状态。其主要包括六个 CSS 状态类，属性值都是布尔类型，具体如表 5.1 所示。

表 5.1　CSS 状态类

状态	为 true 时的 CSS 类	为 false 时的 CSS 类
控件是否被访问过	ng-touched	ng-untouched
控件的值是否变化	ng-dirty	ng-pristine
控件的值是否有效	ng-valid	ng-invalid

表单控件的 CSS 类名会根据表单控件状态变化而变化，表单在无效状态与有效状态切换过程中，需要在视觉上进行区分，那么就需要使用表单控件的 CSS 状态类。

使用表单样式验证当输入有效时控件边框变为绿色，不显示错误提示；当表单输入无效时边框变为红色，显示错误提示，效果如图 5.3 所示。

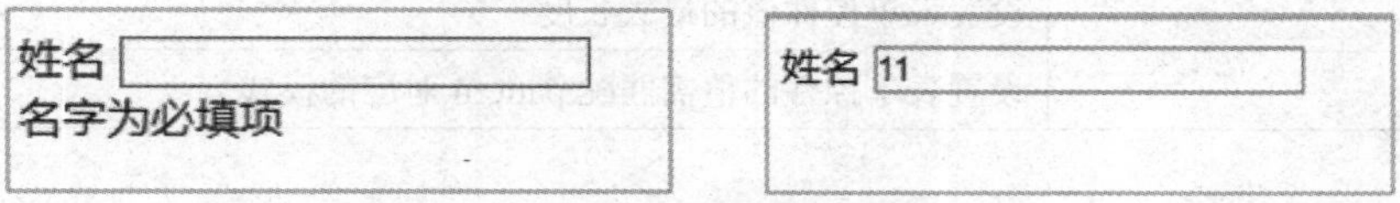

姓名
名字为必填项

姓名 11

图 5.3　表单样式

添加用于视觉反馈的自定义 CSS，代码如下所示。

```
.ng-valid {
  border-left: 5px solid #42A948;
}
.ng-invalid{
  border-left: 5px solid #a94442;
}
```

编辑 HTML 模板用来显示验证信息，代码如下所示。

```
<label > 姓名 </label>
<!--name 是一个模板变量，把他绑定到 ngModel 上，获取 input 输入值 -->
<!--hidden 表示字段有效或者值没有变化时隐藏提示 -->
<input type="text" required [(ngModel)]="model.name" name="name"  #name="ngModel">
<div [hidden]="name.valid"   class="alert alert-danger">
   名字为必填项
</div>
```

3 表单验证

表单验证是用来检测表单的输入值是否满足设定的规则，当用户输入的数据格式不正确时，则将相关状态立即反馈给用户，从而增强用户体验。HTML5 表单内置了相关的基础验证，但这些基础验证使用场景有限。为了能够高效的完成相关表单验证，Angular 提供了内置校验和自定义校验两种方式。

（1）表单内置校验

Angular 表单内置验证与普通的 HTML 表单验证一样，直接在表单控件上添加相应的属性即可，但验证提示语的显示或隐藏是通过指令控制。其支持的内置表单验证器如表 5.2 所示。

表 5.2 表单验证器属性

属性	描述
required	设置表单控件值是非空的
email	设置表单控件值的格式是 email
minlength	设置表单控件值的最小长度
maxlength	设置表单控件值的最大长度
pattern	设置表单控件的值需匹配 pattern 对应的模式

使用表单内置校验代码如下所示。

```
<div class="form-control">
  <label> 用户名 :</label>
```

```
  <!--name 是一个模板变量，把他绑定到 ngModel 上，获取 input 输入值 -->
  <!--hidden 表示字段有效或者值没有变化时隐藏提示 -->
  <input [(ngModel)] = "model.name" type="text" name="username" id="username" required
   #uname="ngModel"/>
  <div [hidden] = 'uname.valid || uname.pristine'>
  用户名必填
  </div>
</div>
```

还可以使用 *ngIf 指令来实现上述效果，代码如下所示。

```
<div class="form-control">
  <label> 用户名 :</label>
<!-- 通过 #uname ="ngModel" 方式获取 input 输入值，然后通过 errors 属性，来获取对应
验证规则 ( 如 required, minlength 等 )-->
 <input [(ngModel)] = "model.name" type="text" name="username" required  #uname="ng-
Model"/>
  <div *ngIf="uname.invalid && (uname.touched || uname.dirty)">
  <div *ngIf="uname.errors.required">
   用户名必填
  </div>
 </div>
</div>
```

（2）自定义验证器

当需要实现复杂的表单校验功能时，表单内置校验则无法满足，这时就需要自定义验证器。自定义验证器是一个返回任意对象的函数，函数传入的参数是 FormControl 类，通过对 FormControl 类的 value 值进行校验处理，返回校验结果。其定义格式为：

```
// fnName 函数名可变
export function fnName(control:FormControl):any{
// 验证内容
}
```

实现自定义验证器验证输入框密码和确认密码是否相同，效果如图 5.4 所示，具体实现步骤如下所示。

第一步：新建自定义验证指令

打开命令窗口输入“ng g directive directives/equalValidator（equalValidator 为自定义验证器名称）”，新建自定义验证指令。

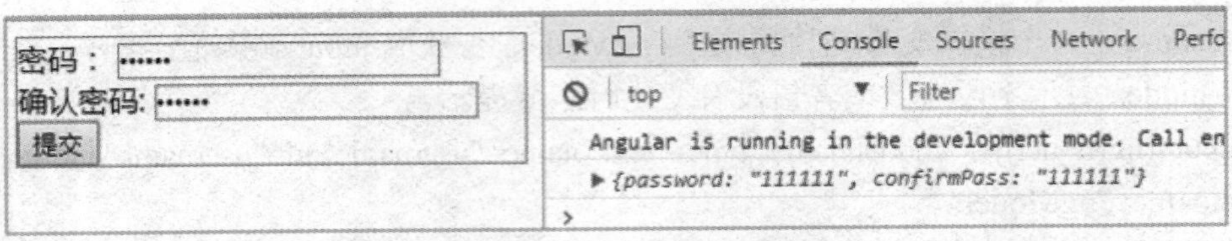

图 5.4 自定义表单验证

第二步：编写验证器

新建 validators.ts 组件（ validators 名字自定义），用于存放一些验证规则。以下为验证输入框密码和确认密码是否相同。代码如 CORE0501 所示。

代码 CORE0501：validator.ts 文件

```
import { FormControl, FormGroup } from '@angular/forms';

export function equalValidator(group: FormGroup):any {
// 获取表单值
  let password: FormControl = group.get("password") as FormControl;
  let cPassword: FormControl = group.get("confirmPass") as FormControl;
  let valid = null;
// 比较表单值
  if(password && cPassword) {
    valid = (password.value === cPassword.value);
  }
  return valid ? null :{"equal": {
    desc:" 密码和确认密码不一致 "
  }};
```

第三步：配置自定义验证器

在第一步新建的 equalValidator 指令文件中，配置一个 providers 提供器，其 provide 值固定为 NG_VALIDATORS，useValue 为定义的验证器方法名。代码如 CORE0502 所示。

代码 CORE0502：equalValidator.ts 文件

```
import { Directive } from '@angular/core';
import {NG_VALIDATORS } from '@angular/forms';
import {equalValidator} from "../validator/validators";

@Directive({
  selector: '[equal]',
  providers :[{
    provide : NG_VALIDATORS,
```

```
      useValue : equalValidator,
      multi: true
   }],
})
export class EqualValidatorDirective {

   constructor() { }
}
```

第四步：编辑 HTML 模板

在模板中引入自定义的指令，代码如 CORE0503 所示。

代码 CORE0503：HTML 文件

```
<form #myForm = "ngForm" (ngSubmit) = "onSubmit(myForm.value)" novalidate>
     <label> 密码：</label>
     <input ngModel type="password" name="password" required  minlength="6"
          #pass="ngModel"/>
     <div *ngIf="pass.invalid && (pass.touched || pass.dirty)">
      <div *ngIf="pass.errors.minlength">
       密码长度不少于 6 位
      </div>
     </div>
    <div>
     <label for="confirmPass"> 确认密码 :</label>
     <input ngModel type="password" name="confirmPass" required/>
    </div>
    <div [hidden] = "!myForm.form.hasError('equal','passwordGroup')">
     两次密码不一致！
    </div>
 <input type="submit" value=" 提交 "/>
</form>
```

第五步：在对应的 ts 文件中，编写 onSubmit() 方法，在控制台打印出表单填入的信息。代码如下 CORE0504 所示。

代码 CORE0504：ts 文件

```
onSubmit(value) {
   console.log(value);
      }
```

提示：当设计表单时，为了界面具有响应式风格，你是否需要将表单设置为响应式？扫描图中二维码，你将得到你想要的内容！

技能点 2　依赖注入的介绍

依赖注入简称为 DI，是一种很重要的程序设计模式，在这种模式下，一个或多个依赖（或服务）被注入到一个独立的对象中，然后成为了该对象的一部分。其有助于应用程序各模块之间的解耦，使代码更容易维护。

依赖注入的核心概念包括注入器（Injector）、提供器（Provider）、依赖（Denpendence）。具体介绍如下所示。

- 注入器：提供了一系列的接口用于创建依赖对象的实例。其实现方法是在构造函数中声明。
- 提供器：用于配置注入器。其实现方法是在 Provider 通过注入器提供的令牌（Token，可能是字符串也可能是类）创建被依赖的对象。
- 依赖：指定了被依赖对象的类型，注入器会根据此类型创建对应的对象。

使用 Angular 实现依赖注入时，首先需要注入器提供令牌（依赖的标识）来获取服务，通常在构造函数里面为参数指定类型，该参数类型就是依赖注入器所需的令牌；Angular 把该令牌传给提供器，提供器根据该令牌创建被依赖的对象。

1　注入器

注入器负责注入组件需要的对象，一般注入器会自动通过组件的构造函数，将组件所需的对象注入到组件中。Angular 中注入器的实现方法是在构造函数中声明。具体代码如下所示。

```
constructor(private 对象 : 类型 ){
}
// 示例代码
constructor(private projectServie:ProjectService){
..}
```

注意：在上面代码中组件声明了一个 projectService 对象，并指定它的类型为 ProjectService（只是一个 Token），Angular 注入器看到上面构造函数声明时，就会在提供器中寻找 Token 为 ProjectService 类型的实例，如果没找到，提供器就会根据该令牌创建一个对象，最终被注入到需要的组件或服务中使用。

2　提供器

在 Angular 中，提供器（Provider）描述初始化标识（Token）所对应的依赖服务，最终被注入到组件或服务中使用。实现方法是在 Provider 通过注入器提供的令牌 (Token，可能是字符串也可能是类) 创建被依赖的对象。

当组件在构造函数中需要依赖一个类时，Angular 首先会在该组件寻找提供器，当组件自身没有提供器时，去该组件的父组件中找，如果父组件中也没有，就去根模块（app.module.ts）中寻找。找到后，通过指定 providers 依赖注入。

提供器可以在所有组件或服务中使用，也可以定义在某一个组件中，只供这一个组件使用。不同地方其作用域规则也不相同，具体作用域规则如下所示：

- 当提供器在根模块（app.module.ts）声明时，对所有组件可见。
- 当在组件声明时只对相应的组件和子组件可见，其他不可见。
- 当在模块和组件声明，且具有相同 Token 时，组件的声明覆盖模块的声明。
- 一般把提供器优先声明在模块中。

（1）在模块（如：根模块）中声明提供器

通常提供器声明在根模块（app.module.ts）中，可供所有组件或服务使用，定义在模块中使用 @NgModule 声明。如在 app.module.ts 的 providers 中注入 ProjectService 服务，代码如下所示。

```
@NgModule({
 imports:[],
 declarations:[],
 providers:[ ProjectService],
bootstrap: [ AppComponent ]
})
```

（2）在组件中声明提供器

提供器也可声明在某一个组件中，只供当前组件使用，定义在组件中使用 @Component 声明。如在组件中注入 ProjectService 服务，代码如下所示。

```
import { Component } from '@angular/core';
import { HeroService } from './hero.service';

@Component({
selector: 'app-heroes',
```

```
providers: [ProjectService],
template: ` <h2>Heroes</h2> `
})
export class HeroesComponent { }
```

注意：provide 声明的是在注入器中的 Token，就是说这两个 Token 是对应的。

在 Angular 中提供器分为以下四种：

（1）useClass

类提供器是最常见的注入类型，其中 provide 配置方法接收两个键（key）：第一个是 provide 键，是可注入对象标识的令牌（Token），用于定位依赖值和声明提供器。第二个是 useClass 键，用于定义对象，指出注入对象。具体代码如下所示。

```
providers: [{provide: 令牌 ,useClass: 注入对象 }]
//provide 指定了提供器的 Token, 与构造函数中声明的类型对应 , useClass 指出注入的对象
// 示例代码
providers: [{provide: ProjectService,useClass: BetterLogger}]
```

注意：在这里把 BetterLogger 对象映射到 ProjectService 令牌上。在这个例子中，类名和令牌名是不匹配的，这是最常见的情况，有时令牌和被注入对象是同名的。

（2）useExisting

别名提供器就是用不同的"名字"注入同一个服务。实现在一个 Provider 中配置多个标识，其对应的对象指向同一个实例。制造一个别名来引用注册过的对象，提供器可以把一个令牌映射到另一个令牌上。实际上，第一个令牌是第二个令牌所对应的服务的一个别名，从而实现了多个依赖、一个对象实例的形式。

例如：有一个 ProjectService 记录消息的服务，同时又开发了有相同接口的新服务 Project1Service，但并不想去替代 ProjectService 服务，为了使新旧服务同时可用，可以用 useClass 实现，示例代码如下所示：

```
// 使用 useClass 实现
providers: [
        {provider: ProjectService, useClass: ProjectService },
        {provider: Project1Service, useClass: ProjectService }
]
```

但是，我们不希望应用中有两个不同的 ProjectService 实例，使用 useExisting 可以多个标识指向同一实例。具体代码如下所示。

```
// 示例代码
// 使用 useExisting 实现
```

```
providers: [
        {provider: ProjectService1, useExisting: ProjectService }
]
```

（3）useValue

由于依赖的对象并不一定都是类，也可以是字符串、常量、对象等其他类型，useValue 提供器可以方便用在全局变量、系统相关参数配置场景中。在创建 Provider 对象时，只需使用 useValue 即可声明一个值。具体代码如下所示。

```
// 标识为 someone 依赖的值为变量 freeMan，在使用之前需要先定义
let freeMan = {
    freeJob: boolen;
    live: () => {return ' 使用值 useValue '}
};

// 定义在组件中是通过 @Component 声明
@Component({
    providers: [
        {provide: 'someone', useValue: freeMan}
    ]
})
```

（4）useFactory

有时依赖对象是不明确且动态变化的，工厂提供器允许根据不同的条件来实例化不同的服务，也就是可以动态创建依赖值。在使用时需要写一个返回任意对象的函数，通过调用函数来新建一个依赖对象，并且依赖对象可以作为提供器。具体代码如下所示。

```
providers:[{
    provide: MyComponent,
// 如果 loggedIn 为真，则注入器会返回一个 MyLoggedComponent 的实例；否则返回
//MyComponent 的实例
    useFactory: ()=> {
        if(loggedIn) {
            return new MyloggedComponent();
        }
        return new MyComponent();
    }
}]
```

技能点 3　依赖注入的应用

1　在组件中注入服务

在组件中使用依赖注入需要以下步骤：

第一步：打开命令窗口，通过 ng g service App（App 为服务名称）命令新建 AppService 服务。

第二步：在组件构造函数中声明服务。

为构造函数提供对应的依赖服务，代码如 CORE0505 所示。

代码 CORE0505：ts 文件

```
export class AppComponent {
    // 在构造函数中声明需要注入的依赖
  constructor(public service: AppService) { }
}
```

第三步：在组件中配置注入器。

在启动组件时，Angular 会读取 @Component 装饰器里的 providers 元数据，它是一个数组，配置了该组件需要使用的所有依赖，Angular 的依赖注入会根据这个数组去创建对应的实例，当前组件和子组件都能共享注入器创建的实例，代码如下所示。

```
@Component({
  selector: 'app-root',
  templateUrl: './app.component.html',
  styleUrls: ['./app.component.css'],
    // 在组件中配置注入器
  providers: [AppService]
})
```

第四步：通过 import 导入被依赖的对象服务（例如导入 AppService 服务），代码如下所示。

```
import { Component } from '@angular/core';
// 导入被依赖对象的服务
import { AppService } from './app.service';
```

2　在服务中注入服务

除了组件依赖服务，服务间依赖也很常见。如：在上面提到的 AppService 服务中，如果希望在异常时记录错误信息，则创建一个记录错误的服务，并把其注入在 AppService 服务中。

创建 LoggerService 服务代码如 CORE0506 所示。

代码 CORE0506：LoggerService.ts

```
import {Injectable} from "@angular/core";

@Injectable()
export class LoggerService{

}
```

把 LoggerService 服务中注入 AppService 服务中。代码如 CORE0507 所示。

代码 CORE0507：AppService.ts

```
import { Injectable } from '@angular/core';
import { AppService } from './app.service';

// 服务与服务之间的依赖必须使用 @Injectable
@Injectable()
export class AppService{
    constructor(private logger: LoggerService) {}
}
```

在根模块的 providers 元数据中注入服务。代码如 CORE0508 所示。

代码 CORE0508：app.module.ts

```
providers: [
    LoggerService,
    AppService
]
```

3　在模块中注入服务

在模块中注入服务和在组件中注入服务的方法是一样的，不同的模块中注入的服务在整个组件中都是可用的，这样增强了模块的扩展性。Angular 在启动程序时会启动一个根模块，并加载其所依赖的其他模块以及生成一个全局的根注入器，由该注入器创建的依赖注入对象在整个应用中可见。代码如 CORE0509 所示。

代码 CORE0509：在模块中注入服务

```
import { NgModule } from '@angular/core';
import { BrowserModule } from '@angular/platform-browser';
```

```
import { HeroService } from './hero.service';
import { Hero1Service } from './hero1.service';
import { Hero2Service }from './hero2.service';

@NgModule({
  imports: [
    BrowserModule,
  ],
  declarations: [
  ],
  providers: [ HeroService ,Hero1Service,Hero2Service],
  bootstrap: [AppComponent]
})
export class AppModule { }
```

（1）在两个模块中使用同样的 Token 注入服务时，因为根注入器只有一个，后初始化的模块服务（BModule）会覆盖前面初始化的模块服务（AModule）。

在 A 模块中注入 projectServie1 服务。代码如下所示。

```
constructor(private projectServie1:ProjectService){ }
```

在 B 模块中注入与 projectServie1 相同的服务（projectServie2 服务）。代码如下所示。

```
constructor(private projectServie2:ProjectService){ }
```

在根模块先后导入 A、B 两个模块，代码如下所示。

```
@NgModule({
imports: [
  AModule,
  BModule
]
})
```

（2）在根模块中注入的服务拥有最高优先级。例如：AModule 模块初始化的服务声明在组件中，BModule 模块初始化的服务是声明在根模块（app.module.ts）时，那么模块使用的都是根模块中注入的服务。

AModule 模块初始化的服务声明在组件中，代码如下所示。

```
import { Component } from '@angular/core';
import { HeroService } from './hero.service';
```

```
@Component({
selector: 'app-heroes',
providers: [ProjectService],
template: ` <h2>Heroes</h2> `
  })
export class HeroesComponent { }
```

BModule 模块初始化的服务是声明在根模块，代码如下所示。

```
@NgModule({
 imports:[],
 declarations:[],
 providers:[ ProjectService],
bootstrap: [ AppComponent ]
})
```

在模块中注入自定义 product 服务实现如图 5.5 所示效果。

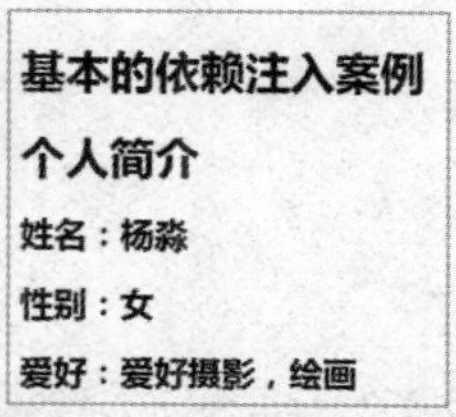

基本的依赖注入案例

个人简介

姓名：杨淼

性别：女

爱好：爱好摄影，绘画

图 5.5　依赖注入

实现图 5.5 效果步骤如下：

第一步：打开命令窗口，使用命令新建组件（product1）、服务，代码如下所示。

```
// 新建组件
ng g component product1
// 新建服务
ng g service shared/product
```

第二步：在 app.component.html 主组件写入 product1 组件。代码如 CORE0510 所示。

```
代码 CORE0510：app.component.html
<div>
  <div>
    <h1> 基本依赖注入样例 </h1>
  </div>
</div>
```

```
<app-product1></app-product1>
```

第三步：在 product1.component.html 编写 HTML 模板。代码如 CORE0511 所示。

代码 CORE0511：HTML 模板

```
<div>
  <h1> 个人简介 </h1>
  <h2> 姓名：{{product.title}}</h2>
  <h2> 性别：{{product.price}}</h2>
  <h2> 爱好：{{product.desc}}</h2>
</div>
```

第四步：在 product.service.ts 文件中定义一个类并定义其属性。代码如 CORE0512 所示。

代码 CORE0512：product.service.ts

```
import { Injectable } from '@angular/core';
// 能把别的服务注入 ProductService 服务中
@Injectable()
export class ProductService {
  constructor() { }

getProduct(): Product{
  return new Product(0," 杨淼 "," 女 "," 爱好摄影，绘画 ")
}
}
// 创建一个类构建字段
export class Product{
  constructor(
     public id:number,
     public title:string,
     public price:number,
     public desc:string
  ){
  }
}
```

第五步：在 product1.component.ts 文件中的构造方法声明要注入的服务，代码如下所示。

```
import { Component, OnInit } from '@angular/core';
import {Product} from "../shared/product.service";
```

```
import {ProductService} from "../shared/product.service";

@Component({
  selector: 'app-product1',
  templateUrl: './product1.component.html',
  styleUrls: ['./product1.component.css']
})
export class Product1Component implements OnInit {
// 声明一个属性来接收服务传来的属性
product:Product;
// 通过构造函数声明需要一个 Token 为 ProductService 类型的服务
  constructor(private productService:ProductService) { }
  ngOnInit() {
    this.product=this.productService.getProduct();
  }
}
```

第六步：在 app.module.ts 中初始化 Token 所对应的依赖服务，代码如下所示。

```
import {ProductService} from "./shared/product.service";
@NgModule({
  providers:[ProductService],
})
export class AppModule { }
```

快来扫一扫！

提示：在学会了使用 Angular 依赖注入之后，你是否想要知道 AngularJS 依赖注入的使用？扫描图中二维码，了解更多详情。

通过下面十一个步骤的操作，实现图 5.2 所示的智慧工厂水监控模块的效果。

第一步：将水监控模块分为头部图标、水压监控图组件以及简介部分，并使用命令创建头部图标、简介组件以及路由文件。

第二步：在命令窗口输入以下命令创建服务，并在 water-monitoring.module.ts 文件创建服务。命令如下所示。

```
ng g service book
```

在 water-monitoring.module.ts 文件中注入服务。代码如下所示。

```
import {BookService} from "./book.service";

@NgModule({
  providers: [BookService],
})
```

第三步：在 water-monitoring.component.html 文件中对模块进行布局，设置各个组件的渲染位置。代码如 CORE0513 所示。

代码 CORE0513：water-monitoring.component.html

```
<!-- 头部图标 -->
<app-water-table></app-water-table>
<h3 class="text-center"></h3>
<div class="row" style="margin-left: 50px">
    <div class="col-md-11">
        <div class="widget green">
            <div class="widget-title">
                <h4 style="color: white"><i class="icon-reorder"></i> 自来水水压监控图 </h4>
            </div>
            <div class="widget-body" >
    <!-- 水压监控图组件 -->
            </div>
        </div>
    </div>
</div>
     <!-- 文字简介 -->
<div class="row" style="margin-left: 50px">
    <div class="col-md-11">
    <app-water-text></app-water-text>
</div>
</div>
```

第四步：在 water-table.component.html 文件中，编写头部图标。头部图标主要是对水质、水位的显示。代码如 CORE0514 所示。设置样式前效果如图 5.6 所示。

代码 CORE0514：water-monitoring.component.html

```
<h1 class="text-center"> 智慧工厂水监控信息显示 </h1>
<div class="infobox-container">
  <div class="infobox infobox-green infobox-small infobox-dark">
        <div class="percentage">
           <span>61%</span>
        </div>
     <div class="infobox-data">
        <div> 目标 </div>
        <div> 完成 </div>
     </div>
 </div>
 <div class="infobox infobox-blue infobox-small infobox-dark">
  <div class="infobox-data">
     <div> 水质 </div>
     <div> 未超标 </div>
  </div>
 </div>
 <div class="infobox infobox-grey infobox-small infobox-dark">
  <div class="infobox-data">
    <div> 水位 </div>
    <div> 未超标 </div>
  </div>
 </div>
</div>
```

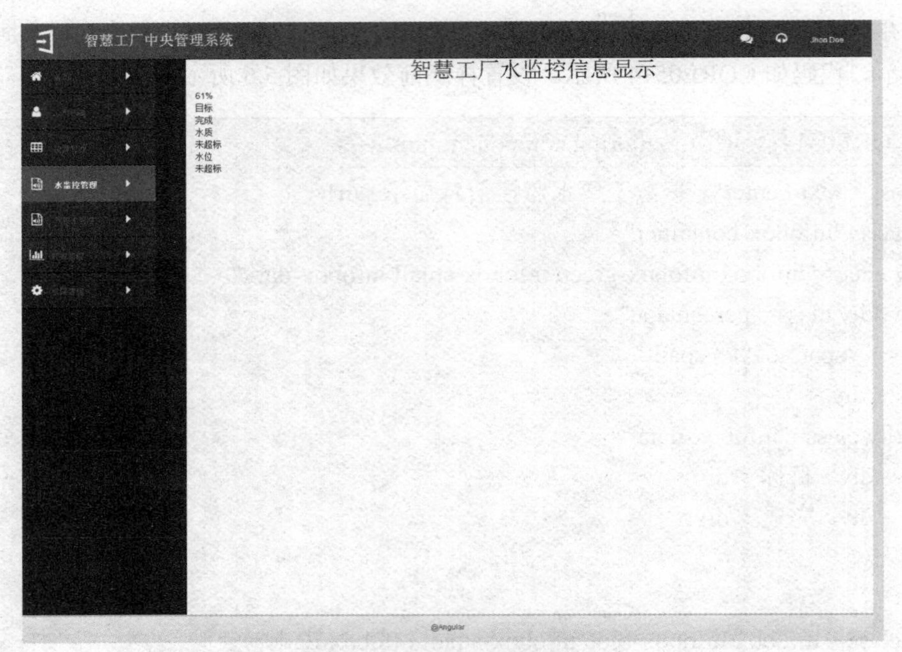

图 5.6　头部图标设置样式前

设置头部图标颜色、字体大小颜色等，代码如 CORE0515 所示。设置样式后效果如图 5.7 所示。

代码 CORE0515：CSS 代码

```
.infobox-container{
text-align:center;
font-size:0
}
.infobox{
  display:inline-block;
  width:210px;
  height:66px;
  color:#555;
  background-color:#FFF;
  box-shadow:none;
  border-radius:0;
  margin:-1px 0 0 -1px;
  padding:8px 3px 6px 9px;
  border:1px dotted;
  border-color:#d8d8d8!important;
  vertical-align:middle;
```

```
    text-align:left;
    position:relative
}
.infobox>.infobox-data{
    display:inline-block;
    border:0;
    border-top-width:0;
    font-size:13px;
    text-align:left;
    line-height:21px;
    min-width:130px;
    padding-left:8px;
    position:relative;
    top:0
}
.infobox>.infobox-data>.infobox-data-number{
    display:block;
    font-size:22px;
    margin:2px 0 4px;
    position:relative;
    text-shadow:1px 1px 0 rgba(0,0,0,0.15)

}
.infobox>.infobox-data>.infobox-text{
    display:block;
    font-size:16px;
    margin:2px 0 4px;
    position:relative;
    text-shadow:none
}
```

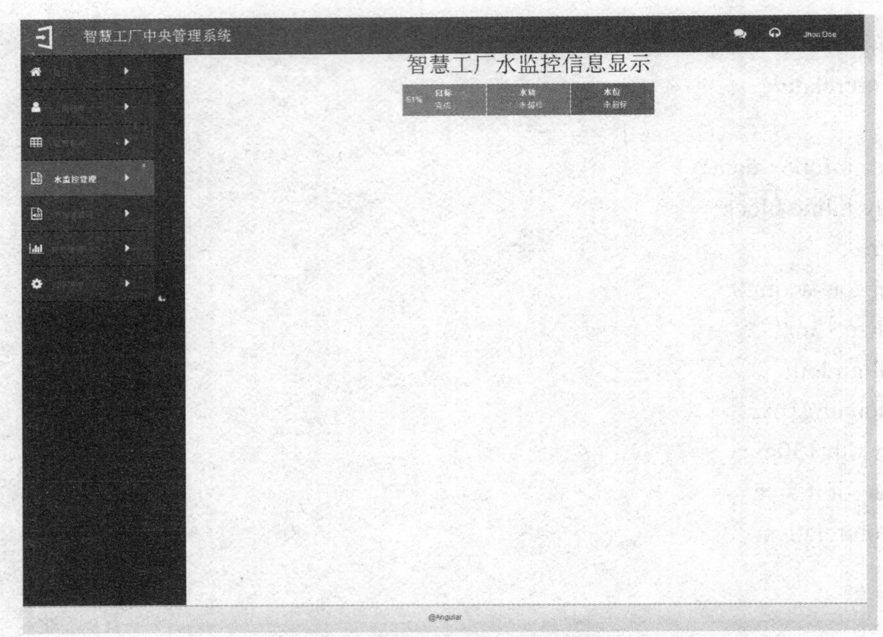

图 5.7 头部图标设置样式后

第五步：安装图表环境。打开命令窗口，使用 npm 安装图表。代码如下所示。

```
npm install ng2-charts --save
```

安装图表包含的库。并在 index.html 文件中引入 Chart.js。代码如下所示。

```
// 安装类库
npm install chart.js –save
// 引入 Chart.js 文件
<script src="node_modules/chart.js/src/chart.js"></script>
```

在 app.module.ts 文件中引入 ChartsModule。代码如下所示。

```
import { ChartsModule } from 'ng2-charts';
imports: [
    ChartsModule
]
```

第六步：在 water-monitoring.component.html 中编写底部水压监控图组件。水压监控图组件包含头部标题、图表部分以及一个更新按钮。代码如 CORE0516 所示。

代码 CORE0516：底部水压监控图组件

```
<div class="row" style="margin-left: 50px">
  <div class="col-md-11">
```

```
    <div class="widget green">
      <div class="widget-title">
        <h4 style="color: white"> 自来水水压监控图 </h4>
      </div>
      <div class="widget-body" >
        <div class="text-center">
            <div style="display: block;height: 5%">
        <!-- lineChartData 查看关于 label 图例和工具提示中出现的数据集
        lineChartLabels x 轴标签
        lineChartColors 设置图表颜色
        lineChartOptions 设置图表选项
        lineChartType 设置图表类型为柱状图 -->
            <canvas baseChart
                    [datasets]="lineChartData"
                    [labels]="lineChartLabels"
                    [colors]="lineChartColors"
                    [chartType]="lineChartType">
            </canvas>
          </div>
          <button (click)="randomize()"> 更新 </button>
        </div>
      </div>
    </div>
  </div>
</div>
```

在 water-monitoring.component.ts 文件中设置图表的类型、x 轴数据、图表选项等。如 CORE0517 所示。效果如图 5.8 所示。

代码 CORE0517：water-monitoring.component.ts

```
public lineChartColors:Array<any> = [
  {
    backgroundColor: '#5fb3e0',
  }
]
                                                                                                    ;

//x 轴
public lineChartLabels:string[] = ['10 月 1 日 ', '10 月 2 日 ', '10 月 3 日 ', '10 月 4 日 ', '10 月 5 日 ', '10 月 5 日 ', '10 月 7 日 ', '10 月 7 日 ', '10 月 7 日 ', '10 月 7 日 ', '10 月 7 日 '];
```

```
public lineChartType:string = 'line';
public barChartLegend:boolean = true;
// 图表更新事件
public randomize():void {
      let data = [
         Math.round(Math.random() * 100),
         59,
         80,
         (Math.random() * 100),
         55,
         (Math.random() * 100),
         40];
// parse 用于从一个字符串中解析出 JSON 对象 ,stringify() 用于从一个对象解析出字符串
      let clone = JSON.parse(JSON.stringify(this.barChartData));
      clone[0].data = data;
      this.barChartData = clone;
 }
```

第七步：新建 ts 文件，声明一个类并定义两个变量，用于存放监控图组件的数据。

```
export class chartdata {
  data1: string;
  data2: string;
  constructor() {
  }
}
```

第八步：编写静态的本地数据。在 src 的 assets 目录下新建 data 文件夹，并在 data 目录下新建 chartdata.json 文件（名字可以修改），写入数据。代码如 CORE0518 所示。

代码 CORE0518：静态的本地数据

```
[
  {
    "data1":30,
  },
  {
    "data1":70,
  },
// 省略部分代码
 ]
```

第九步：在新建的 book.service.ts 文件中注入 Http 服务、rxjs 操作符（是一个基于可观测数据流在异步编程应用中的库），以及导入新建的 chartdata.ts 文件，代码如下所示。

```
import { Injectable } from '@angular/core';
import { Http, Response } from '@angular/http';
import { Observable } from 'rxjs';
import 'rxjs/add/operator/map';
import 'rxjs/add/operator/toPromise';
// 导入新建的 chartdata.ts 文件
import { Chartdata} from './ chartdata';

@Injectable()
export class BookService {
// 注入 Http 服务
    constructor(private http:Http) { }
}
```

在 book.service.ts 中引入 JSON 文件，代码如下所示。

```
export class BookService {
    url = "http://localhost:4200/data/books.json";
    }
```

使用 http.get() 方法获取 URL 地址中的数据。代码如下所示。

```
getBooksWithObservable(): Observable<Book[]> {
// 获取 URL 地址中的数据
  return this.http.get(this.url)
    .map(this.extractData)
  }
```

添加 extractData() 方法。将 Response 对象转换成 JSON 串并返回，代码如下所示。

```
private extractData(res: Response) {
   let body = res.json();
       return body;
  }
```

第十步：在 water-monitoring.component.ts 中获取 JSON 文件，HTTP 请求使用 subscribe() 方法调用 data 里的数据，并使用 for 循环输出到定义好的 arr、arr1 变量数组中。代码如 CORE0519 所示。

代码 CORE0519：water-monitoring.component.ts

```
import { Component, OnInit } from '@angular/core';
import { Observable } from 'rxjs';
import {Book} from "./book";
import {BookService} from "./book.service";
@Component({
  selector: 'app-water-monitoring',
  templateUrl: './water-monitoring.component.html',
})
export class WaterMonitoringComponent implements OnInit {
  observableBooks: Observable<Book[]>;
  books: Book[];
// 定义 arr、arr1 变量数组，用于接收 JSON 中的数据
  arr:any;
  arr1:any;
  public lineChartData:any[] = [
    {data: " ", label: ' 自来水 '}
  ];
// 注入 bookService 服务
  constructor(private bookService: BookService) { }
  ngOnInit(): void {
    this.observableBooks = this.bookService.getBooksWithObservable();
    this.observableBooks.subscribe(
// 使用 subscribe() 方法调用 data 里的数据，并循环输出到定义好的 arr、arr1 数组中
        books => {
          console.log(books);
          var ss=books;
          alert(ss.length)
          this.arr=[];
          this.arr1=[];
          for (var i=0;i<ss.length;i++){
            this.arr.push(ss[i].data1);
            this.arr1.push(ss[i].data2);
          }
          console.log(this.arr)
 this.barChartData=[{data: this.arr, label: ' 深井水 '},
                    {data: this.arr1, label: ' 自来水 '}];]
```

```
            },
            error => this.errorMessage = <any>error);
    }
}
```

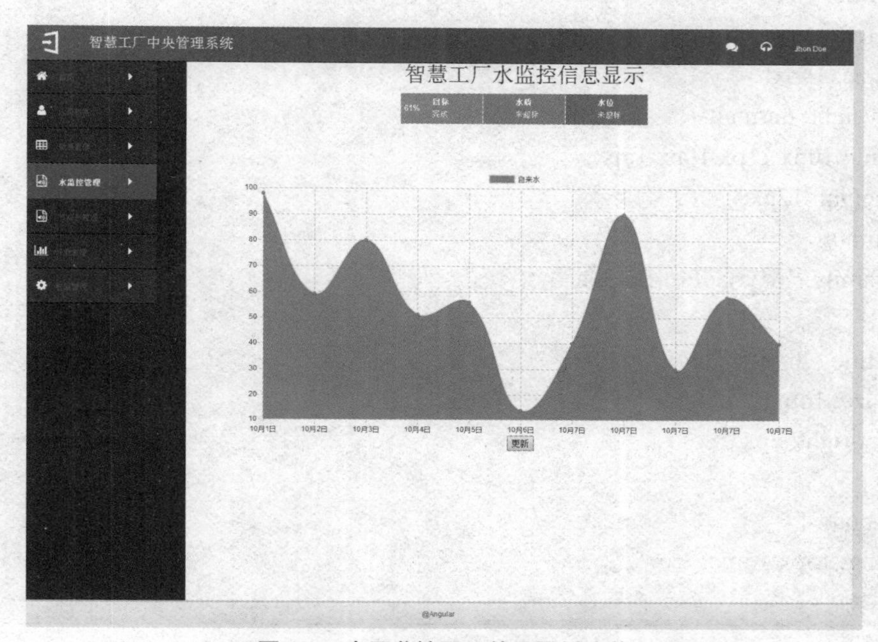

图 5.8　水压监控图组件设置样式前

设置水压监控图组件大小、标题背景为绿色等，代码如 CORE0520 所示。设置样式后效果如图 5.9 所示。

代码 CORE0520：CSS 代码

```
.widget {
  background:#fff;
  clear: both;
  margin-bottom: 20px;
  margin-top: 0;
  border: 1px solid #404040;
}
.widget-title {
  background: #404040;
}
.widget.green {
```

```
    border: 1px solid #74B749;
}
.widget.green .widget-title {
    background: #74B749;
}
.widget-title > h4 {
    float: left;
    font-size: 14px;
    font-weight: normal;
    padding: 10px 11px 10px 15px;
    line-height: 12px;
    margin: 0;
    font-family: 'MyriadPro-Regular';
}
.widget-title > h4 i {
    font-size: 14px;
    margin-right: 2px;
}
.widget-body {
    padding: 15px 15px;
}
```

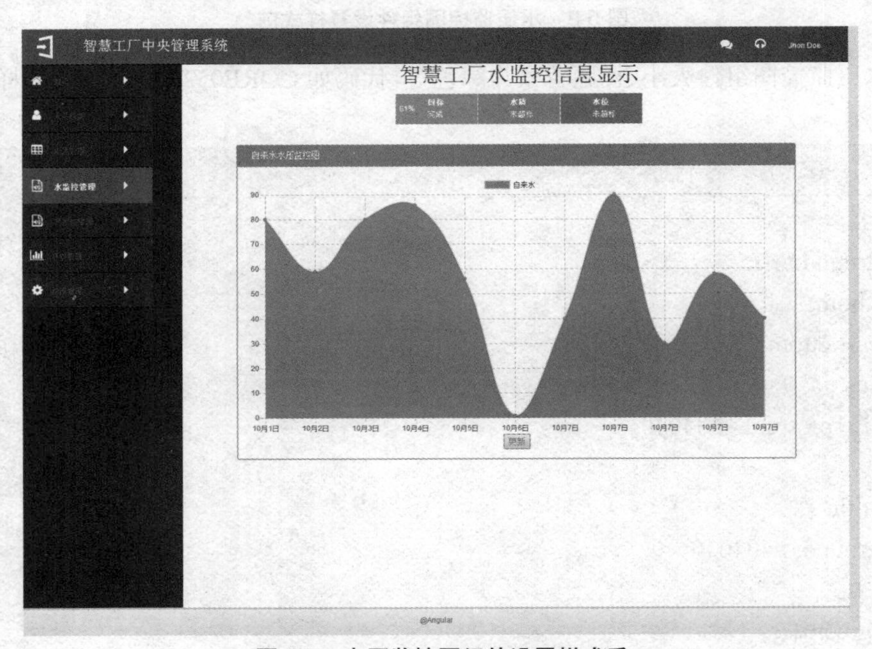

图 5.9　水压监控图组件设置样式后

第十一步：在 water-text.component.html 文件中，编写简介部分。使用栅格系统布局，分为左侧文字简介以及右侧图片部分。代码如 CORE0521 所示。效果如图 5.2 所示。

代码 CORE0521：water-text.component.html

```
<div class="row-fluid">
  <div class="col-lg-7">
    <h4> 用水量监控 </h4>
    <p style="text-indent: 2em;line-height: 30px">
      水资源监控系统，是一种软件与硬件结合的自动化网络式管理系统。它是在水源或用水单位设备上安装一个水资源测控器，实现对水表流量、水井水位、管网压力及对用户水泵的电流、电压的采集，以及对水泵的启停、电动阀的开闭等控制，通过有线或无线通讯方式与水利局水资源管理中心计算机联网，实时对各用水单位进行监管和控制。相关的水表流量、水井水位、
    </p>
  </div>
  <div class="col-lg-5">
    <p>
      <img src="assets/img/1.png" style="height:100%;width: 100%">
    </p>
  </div>
</div>
```

至此，智慧工厂水监控模块制作完成。

通过对智慧工厂水监控模块的学习，对水监控模块中依赖注入等所需知识具有初步了解，掌握提供器与注入器的使用方法。了解注入器作用域规则，具有在组件、服务等中使用依赖注入的能力，为不同组件中注入服务打好基础。

injector	注入器
provider	提供器
denpendence	依赖
key	键
token	令牌
validators	验证器

required 需要
pattern 模式
valid 有效的

一、选择题

1. 下面对依赖注入说法错误的是（ ）。
（A）注入器（Injector）：提供了一系列的接口用于创建依赖对象的实例
（B）提供器（Provider）：用于配置注入器，注入器通过它来创建被依赖对象的实例
（C）依赖（Denpendence）：指定了被依赖对象的类型，注入器会根据此类型创建对应的对象
（D）Angular 实现依赖注入顺序是：注入器→依赖注入→提供器
2. 下面对提供器规则说法错误的是（ ）。
（A）当提供器声明在模块时仅对当前组件可见
（B）当声明在组件时只对相应的组件和子组件可见，其他不可见
（C）当声明在模块和声明在组件的有相同 Token 时，声明在组件的覆盖声明在模块的提供器
（D）一般把提供器优先声明在模块中
3. 下面对提供器说法正确的是（ ）。
（A）类提供器是最常见的注入类型，其中 provide 配置方法接收一个键（key）
（B）别名提供器就是用不同的"名字"注入不同服务中
（C）在创建 Provider 对象的时候，只需要使用 useValue 就可以声明一个值
（D）工厂提供器不允许根据不同的条件来实例化不同的服务
4. 在组件使用依赖注入需要步骤错误的是（ ）。
（A）通过 import 导入被依赖的对象服务
（B）在组件中配置注入器
（C）在组件中配置提供器
（D）在组件构造函数中声明需要注入的依赖
5. 下面对在模块中注入服务错误的是（ ）。
（A）在两个模块中使用同样的 Token 注入同一服务后面导入的模块中的服务会覆盖前面导入的服务
（B）在 AModule 中注入的服务，不同于 BMoudle 中提供的实例
（C）在根模块中注入的服务拥有最高优先级
（D）当 BModule 模块是声明在根模块时，那么两个模块使用的都是根模块中注入的服务

二、填空题

1. 依赖注入是一种 ________ 模式。
2. 依赖注入的核心是 ________、________、________。

3. 类提供器是最常见的注入类型，其中 provide 配置方法接收两个键（key）：________ 和 ________。

4. 注入器是将提供器实例好的对象注入到 ________ 中。

5.Angular 实现依赖注入时，首先需要 ________，方便创建或返回服务，然后通过 ________ 为服务请求一个注入器。

三、上机题

使用 Angular 编写符合以下要求的页面。

要求：用户信息输入的表单页面。在手机号的元素上添加一个验证手机号的验证器。如果手机号验证失败，就显示一个错误提示，页面如下：

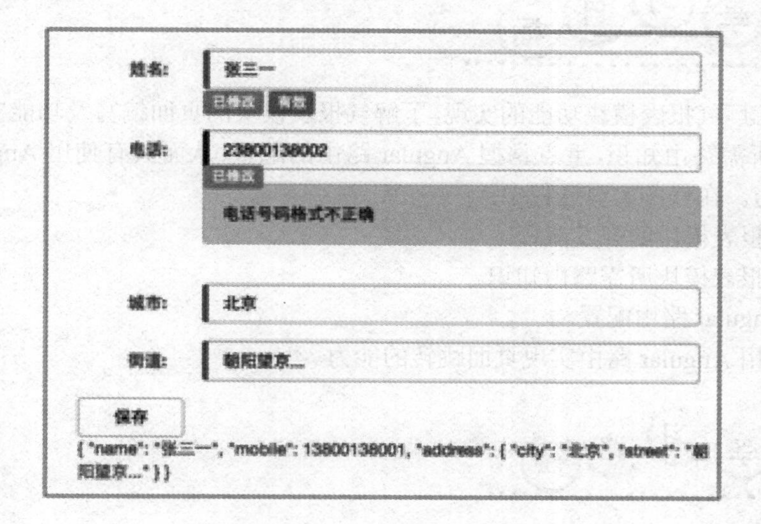

项目六 智慧工厂气报表模块

通过智慧工厂气报表模块功能的实现，了解气报表模块的页面编写及功能实现的流程，学习气报表模块所需路由知识，重点掌握 Angular 路由的配置，从而具有使用 Angular 路由实现页面跳转的能力。在任务实现过程中：

- 了解气报表模块的开发流程。
- 学习气报表模块所需路由知识。
- 掌握 Angular 路由配置。
- 具有使用 Angular 路由实现页面跳转的能力。

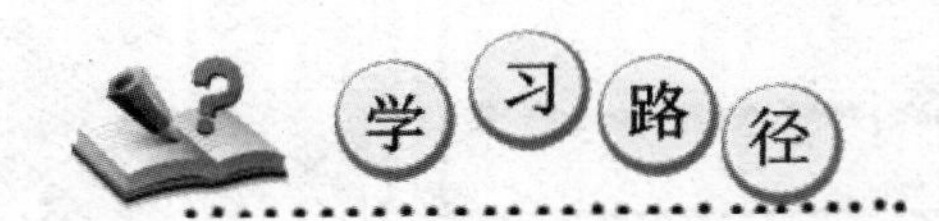

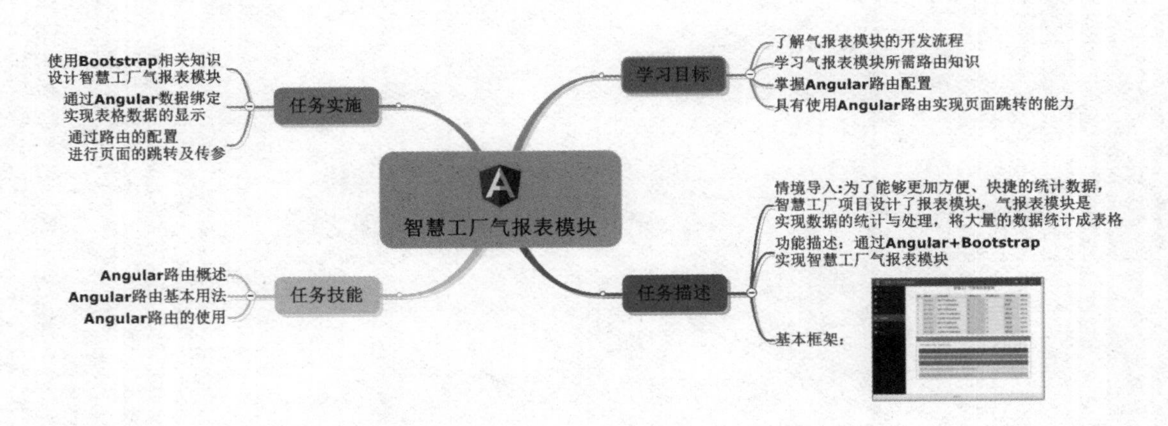

【情境导入】

为了能够更加方便、快捷的统计气体质量数据，智慧工厂项目设计了气报表模块，气报表模块将数据以表格的形式展现。根据气报表模块可以快速查出某一时刻气体报警的详细信息。本项目主要是通过实现智慧工厂的气报表模块来学习 Angular 路由的配置与使用。

【功能描述】

使用 Bootstrap+Angular 实现智慧工厂气报表模块：

- 使用 Bootstrap 相关知识设计智慧工厂气报表模块。
- 通过 Angular 数据绑定实现表格数据的显示。
- 通过路由的配置进行页面的跳转及传参。

【基本框架】

基本框架如图 6.1 所示，通过本项目的学习，能将图 6.1 所示的框架图转换成智慧工厂气报表模块，效果图如图 6.2 所示。

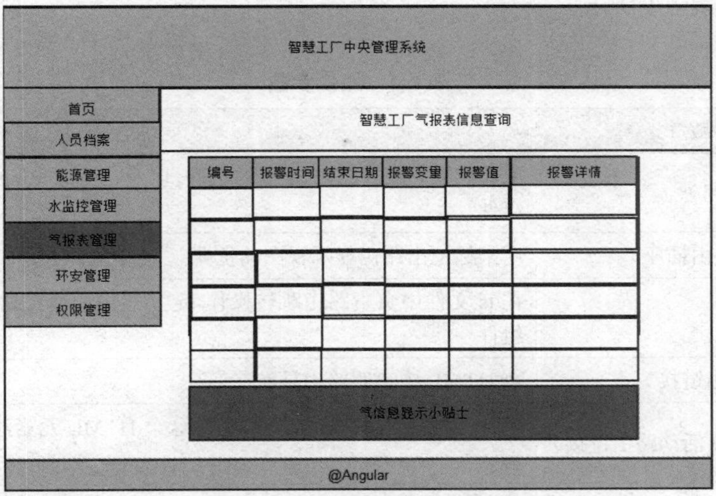

图 6.1　气报表模块框架图

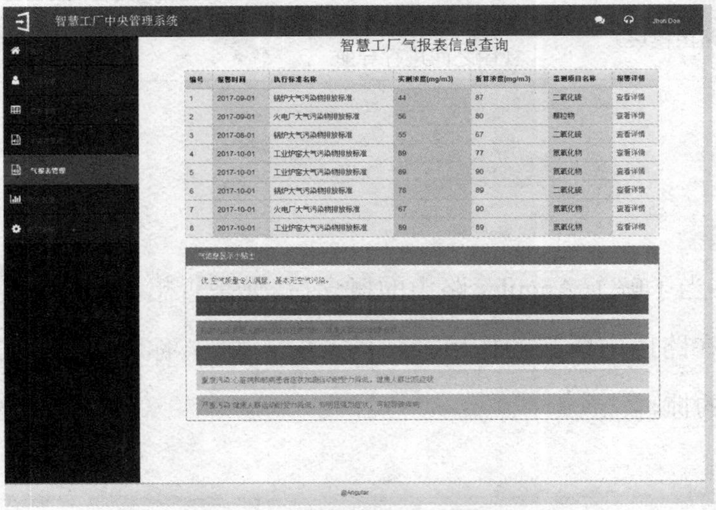

图 6.2　气报表模块效果图

技能点 1 Angular 路由概述

一个 Angular 应用通常可以包含多个模块，每个模块中也可包含多个组件，Angular 路由可以实现不同模块的组件协同工作。实现从一个视图导航到另一个视图，且可以附加可选参数传递给组件，通过不同的 URL 跳转到不同的页面 (HTML)。其实现需要配置路由，配置 Angular 路由参数如表 6.1 所示。

表 6.1 路由参数

参数	说明
Routes（路由数组）	配置路由，保存 URL 对应的组件，以及在哪个 RouterOutlet 中展现组件
RouterOutlet（路由插座）	用来标记出路由显示视图的位置
Router（路由）	在 ts 文件中负责路由跳转操作，定义路由如何根据 URL 来导航到组件
RouterLink（路由链接）	在 HTML 中声明路由导航
routerLinkActive（活动路由链接）	当前激活路由的样式，该指令为这个 HTML 元素添加或移除 CSS 类
redirectTo	重定向
ActivedRoute（激活的路由）	当前激活的路由对象，保存当前路由的信息，如路由地址、参数等
RouterModule（路由模块）	一个独立的 Angular 模块，用于提供所需的服务，以及用来在应用视图之间进行导航

快来扫一扫！

提示：当了解了 Angular 路由的概念后，你是否想知道路由器在路由中所起的作用。扫描图中二维码，你将获取意外的惊喜。

技能点 2　Angular 路由基本用法

当页面中的元素绑定一个路由时，用户点击元素就会导航到相应的视图。例如，当用户点击按钮或从下拉框中选取元素时，可以通过路由进行导航。要做到路由导航，需要使用以下主要参数来配置 Angular 路由：Routes（定义路由配置）、RouterOutlet（标记放置路由的内容位置）、RouterLink 指令（创建路由链接）。路由的基本用法如下。

1　导入 RouterModule

第一步：在 app.module.ts 文件中，从 @angular/router 库中入 RouterModule，代码如下所示。

```
import { NgModule } from '@angular/core';
import { BrowserModule } from '@angular/platform-browser';
import { RouterModule } from '@angular/router';

import { AppComponent } from './app.component';

@NgModule({
  imports: [
    BrowserModule,
    RouterModule
  ],
  bootstrap: [
    AppComponent
  ],
  declarations: [
    AppComponent
  ]
})
export class AppModule {}
```

2　配置路由

为了在应用中使用路由，需要配置路由。使用 RouterModule.forRoot() 来为应用程序提供使用路由必需的依赖。RouterModule 对象提供了两个静态的方法：forRoot() 和 forChild() 来配置路由信息。

（1）RouterModule.forRoot()

在 NgModule 中的 imports 数组里使用 RouterModule.forRoot() 方法来配置路由。该方法提供了路由需要的路由服务和指令。具体代码如下所示。

```
import { Routes, RouterModule } from '@angular/router';
//const 定义路由的配置信息，然后把它作为参数调用 RouterModule.forRoot() 方法
// 每个带路由的 Angular 应用都有一个 Router 服务的单例对象
export const ROUTES: Routes = [];
@NgModule({
  imports: [
    BrowserModule,
    RouterModule.forRoot(ROUTES)
  ],
})
export class AppModule {}
```

（2）RouterModule.forChild()

forChild() 方法与 forRoot() 方法类似，但它只能应用在特性模块（特性模块通过自己提供的服务、组件、指令、管道来与根模块等其他模块协同工作）中。根模块中使用 forRoot()，子模块中使用 forChild()，可以在特性模块中定义模块特有的路由信息，并在必要的时候将其导入根模块。具体代码如下所示。

```
import { NgModule } from '@angular/core';
import { CommonModule } from '@angular/common';
import { Routes, RouterModule } from '@angular/router';

export const ROUTES: Routes = [];
@NgModule({
  imports: [
    CommonModule,
    RouterModule.forChild(ROUTES)
  ],
})
export class ChildModule {}
```

配置路由需要用到 Routers 数组，数组的每一个元素即为一个路由配置项。其配置需要 path、component 两个属性，path 属性定义路由的匹配路径（path 不能用斜线开头），component 属性定义路由匹配时需要加载的组件。具体配置路由数组方式有以下几种：

- 重定向路由。它通常被放在最前面，表示应用的默认路由。
- 在 path 中传参。path 属性定义路由的匹配路径后添加携带的参数。

● 在 path 中存储数据。使用 data 属性存储与此相关联的任意数据。例如存储页面标题等数据。

● Path 为 **（通配符）。通配符路由一般放在最后面，如果当前 URL 无法匹配路由数组中任何一个路由时，路由就会匹配通配符后面组件。

配置路由代码如 CORE0601 所示。

代码 CORE0601：ts 文件

```
import { Routes } from '@angular/router';
// 配置路由数组
const routes: Routes = [
{ path: 'crisis-center', component: CrisisListComponent },
{ path: ' product /:id', component: ProductDetailComponent },
{ path: ' product ', component: ProductListComponent,
data: { title: 'Heroes List' } },
// 当 URL 路径是空时，这个默认路由会重定向到 URL 为 /product
{ path: ' ', redirectTo: '/product', pathMatch: 'full' },
// 可用于显示"404 - Not Found"页面或自动重定向到其他路由
{ path: '**', component: PageNotFoundComponent }
    ];
```

3 调用 RouterOutlet 指令

RouterOutlet 指令是用来标记路由显示视图的位置，指定在页面的哪一位置渲染路由的内容。如：当浏览器中的 URL 变为 /product 时，路由就会匹配到 path 为 product 的 Route，并在主视图中的 RouterOutlet 中显示 Product 详情组件，代码如下所示。

```
<router-outlet></router-outlet>
// 此处呈现 component
```

4 调用 routerLink 指令

现在配置好路由并找到其渲染位置，那么如何才能让 Angular 导航到一个指定路由呢？可以尝试使用纯 HTML（使用 href 直接链接到路由），但是如果这样做，点击这个链接将触发页面重载，而在开发单页应用时是不允许的。

要解决这个问题，Angular 提供了一个方案，可以在不重载页面的情况下链接路由：使用 routerLink 指令。当在标签上绑定 routerLink 指令，用户通过点击标签，来进行导航，代码如 CORE0602 所示。

代码 CORE0602：HTML

```
<nav>
<ul>
  <li><a [routerLink]="['home']">Home</a></li>
  <li><a [routerLink]="['about']">About</a></li>
</ul>
  </nav>
```

路由的基本用法效果如图 6.3 所示。

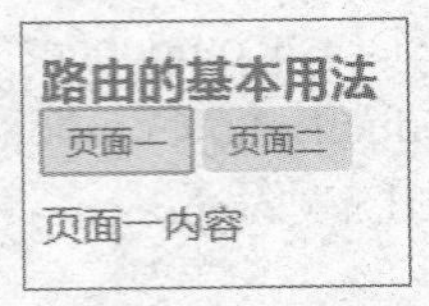

图 6.3　基本路由

为了实现图 6.3 效果，打开命令窗口使用命令新建 dashboard、heroes 两个组件，在 app.component.html 中设置页面布局，使用 routerLink 指令控制导航。代码如 CORE0603 所示。

代码 CORE0603：app.component.html

```
<h1> 路由的基本用法 </h1>
<nav>
  <a routerLink="/dashboard" routerLinkActive="active"> 页面一 </a>
  <a routerLink="/heroes" routerLinkActive="active"> 页面二 </a>
</nav>
<router-outlet></router-outlet>
```

在 dashboard.component.html 页面使用 <div> 标签显示页面一内容。代码如 CORE0604 所示。

代码 CORE0604：dashboard.component.html

```
<div class="grid grid-pad">
页面一内容
</div>
```

在 heroes.component.html 显示页面二内容，代码如 CORE0605 所示。

代码 CORE0605：heroes.component.html

```
<div>
页面二内容
</div>
```

创建一组类型为 Routes 的对象数组，并用它来声明路由配置。路由配置文件代码如 CORE0606 所示。

代码 CORE0606：app-routing.module.ts

```
import { NgModule }       from '@angular/core';
import { RouterModule, Routes } from '@angular/router';

import { DashboardComponent }   from './dashboard.component';
import { HeroesComponent }      from './heroes.component';

const routes: Routes = [
  { path: ' ', redirectTo: '/dashboard', pathMatch: 'full' },
  { path: 'dashboard',  component: DashboardComponent },
  { path: 'heroes',component: HeroesComponent }
];

@NgModule({
  imports: [ RouterModule.forRoot(routes) ],
  exports: [ RouterModule ]
})
export class AppRoutingModule {}
```

快来扫一扫！

提示：学习了 Angular 路由的基本用法后，你有没有想过路由的处理流程？想了解更多的路由使用流程，快来扫我吧！

技能点 3　Angular 路由的使用

1　路由传参

Angular 路由允许通过 URL 向组件传递数据，例如，在一个列表组件中，点击每一项均可加载该示例详情页，这时需要用到路由传参。Angular 路由获取 URL 参数具有两种方式：普通方式传递参数和在路由路径 path 中传递参数。具体介绍如下所示。

（1）普通方式传递参数

普通方式传递参数是通过解析 URL 的 queryParams 属性来获取参数，且每一个配置项都可拥有多个参数。普通方式传递参数具体步骤如下：

第一步：在 HTML 模板中添加传递参数。在基本路由的基础上，添加属性 [queryParams]，使其携带参数。代码如 CORE0607 所示。

代码 CORE0607：app.component.html

```
<button routerLink="/page1" [queryParams]="{bookname:'《初识 Angular》'}" routerLink-Active="active"> 页面一 </button>
```

第二步：在 ts 文件中，创建一个变量来接收传递过来的参数，使用 ActivatedRoute 服务保存当前路由的信息，如路由地址、参数等，并使用参数快照（snapshot 属性）接收路由传递过来的参数。代码如 CORE0608 所示。

代码 CORE0608：page1.component.ts

```
import { Component, OnInit } from '@angular/core';
import {ActivatedRoute, Params} from '@angular/router';
@Component({
  selector: 'app-page1',
  templateUrl: './page1.component.html',
  styleUrls: ['./page1.component.css']
})
export class Page1Component implements OnInit {
  private bookname: string;

  constructor(private activatedRout: ActivatedRoute) {
  }

  ngOnInit() {
    // 普通方式传参的快照获取方式
    this.bookname = this.activatedRout.snapshot. queryParams ['bookname'];
  }
}
```

（2）在 path 中传递参数

顾名思义，其传递方式是通过解析 URL 的 path 来获取参数，path 属性值可以看成两部分，前面是路径，后面是携带的参数。在 path 中传递参数具体步骤如下：

第一步：在 app-routing.module.ts 中修改 path 属性值使其携带参数。代码如 CORE0609 所示。

代码 CORE0609：app-routing.module.ts

```
import { NgModule } from '@angular/core';
import { Routes, RouterModule } from '@angular/router';
import {Page2Component} from "app/page2/page2.component";

const routes : Routes = [
  {path: 'page2/:id',component:Page2Component},
  {path: '**',component:Page404Component},
];

@NgModule({
  imports: [RouterModule.forRoot(routes,{useHash:true})],
  exports: [RouterModule],
  providers: []
})
export class AppRoutingModule { }
```

第二步：在 HTML 模板添加传递参数。

与普通传参类似，使用 path 传参同样是通过 routerLink 指令来实现，routerLink 有两个参数，第一个值为路由的跳转路径，第二个值为路由携带的参数值。代码如 CORE0610 所示。

代码 CORE0610：app.component.html

```
// 当点击模板上标签时则会带着一个 ID 等于 1 的参数跳转
<a [routerLink]="['/page2',1]" routerLinkActive="active"> 列表二 </a>
```

第三步：使用参数快照（snapshot 属性）获取参数值。其实现需要导入 ActivatedRoute 对象，该对象包含一个快照，记录从当前 URL 中解析出来的所有 path 参数。将普通传参 queryParams 属性改为 params。代码如 CORE0611 所示。

代码 CORE0611：page2.component.ts

```
import { ActivateRoute } from '@angular/router';
// 声明一个属性接收传递过来的数据
public data: any;
// 在构造函数里注入 ActivatedRoute 对象
  constructor( public activeRoute:ActivateRoute ) { };
ngOnInit(){
// 使用参数快照获取参数 , route.snapshot 提供了路由参数的初始值。可以通过它来直
// 接访问参数
  this.data = this.route.snapshot.params'id';
```

```
};
```

使用路由传参效果如图 6.4 所示。

图 6.4 路由传参

为了实现图 6.4 效果，新建父路由组件。打开命令窗口，输入以下命令。

```
ng g component home
ng g component books
```

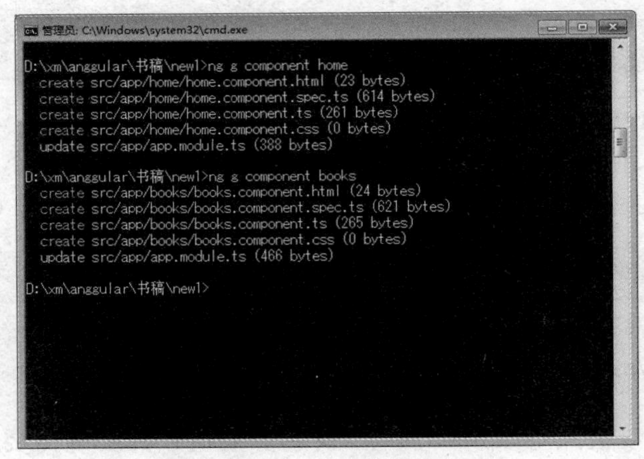

图 6.5 新建主组件

在路由配置文件中定义传入参数的名称。代码如 CORE0612 所示。

代码 CORE0612：app-routing.module.ts

```
const routes: Routes = [
  {path: 'home', component: HomeComponent},
  {
    path: 'books/:bookname',
    component: BooksComponent,
  },
]
```

编写 app.component.html 组件，渲染新建组件位置。代码如 CORE0613 所示。

代码 CORE0613：app.component.html

```
<button [routerLink]="['/home']"> 页面一 </button>
<button [routerLink]="['/books','《活着》']"> 页面二 </button>
<router-outlet></router-outlet>
```

对应的 ts 代码如 CORE0614 所示。

代码 CORE0614：ts 文件

```
import {Component, OnInit} from '@angular/core';
import {ActivatedRoute} from '@angular/router';

@Component({
  selector: 'app-books',
  templateUrl: './books.component.html',
  styleUrls: ['./books.component.css']
})
export class BooksComponent implements OnInit {
  public book: Book;
  constructor(private activatedRout: ActivatedRoute) {
  }

  ngOnInit() {
  this.book = this.activatedRout.snapshot.params['bookname'];
  }

}
// 定义书的属性
export class Book {
  public bookname: string;

  constructor(bookname: string) {
    this.bookname = bookname;
  }
}
```

2　重定向路由

重定向路由主要作用是当访问一个特定地址时，将其重定向到另一个指定的地址。添加重定向路由时需要指定 redirectTo 属性来定向到目标路由，同时需要 pathMatch 属性来指定路由匹配 URL 路径的方式，匹配方式有 prefix（字符串默认为前缀匹配）和 full（完全匹配）两种。重定向路由的配置如下。

```
{path:' 要定向的路径 ', redirectTo:' 要定向到的目标路由 ', pathMatch: ' 匹配方式 '}
```

使用重定向路由效果如图 6.6 所示。

图 6.6 重定向路由效果

为了实现图 6.6 效果，在 app-routing.module.ts 文件中配置路由，虽然指定了当路径为 home 时才指向 HomeComponent 组件，但如果希望在访问根路径时直接展示 HomeComponent 的内容，重定向路由可以实现。代码如 CORE0615 所示。

代码 CORE0615：app-routing.module.ts

```
const routes : Routes = [
  {path: ' ', redirectTo: '/home', pathMatch: 'full'},
  { path: 'books/:bookname', component: HomeComponent},
  { path: 'books',component: BooksComponent,},
    ];
```

3 子路由

子路由是嵌套在主路由里，由 children 表明这是子路由且可以无限循环嵌套。Angular 允许一个路由组件嵌套到另一个路由组件中，从而建立起路由的多级嵌套关系。Angular 中定义的子路由继承父路由的路径，且其加载位置在父路由所定义的 RouterOutlet 中。使用子路由效果如图 6.7 所示。

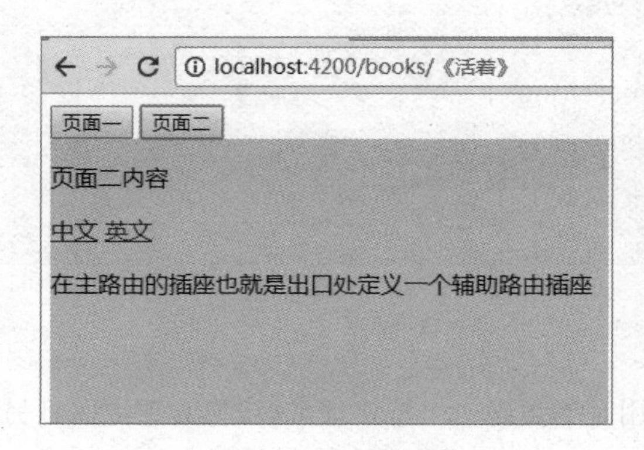

图 6.7 子路由效果

实现上图步骤如下所示。

第一步：新建子路由组件。打开命令窗口，输入以下命令。

```
ng g component introduction_cn
ng g component introduction_en
```

```
管理员: C:\Windows\system32\cmd.exe
D:\xm\anggular\书稿\new1>ng g component introduction_cn
  create src/app/introduction-cn/introduction-cn.component.html (34 bytes)
  create src/app/introduction-cn/introduction-cn.component.spec.ts (685 bytes)
  create src/app/introduction-cn/introduction-cn.component.ts (304 bytes)
  create src/app/introduction-cn/introduction-cn.component.css (0 bytes)
  update src/app/app.module.ts (582 bytes)

D:\xm\anggular\书稿\new1>ng g component introduction_en
  create src/app/introduction-en/introduction-en.component.html (34 bytes)
  create src/app/introduction-en/introduction-en.component.spec.ts (685 bytes)
  create src/app/introduction-en/introduction-en.component.ts (304 bytes)
  create src/app/introduction-en/introduction-en.component.css (0 bytes)
  update src/app/app.module.ts (698 bytes)

D:\xm\anggular\书稿\new1>
```

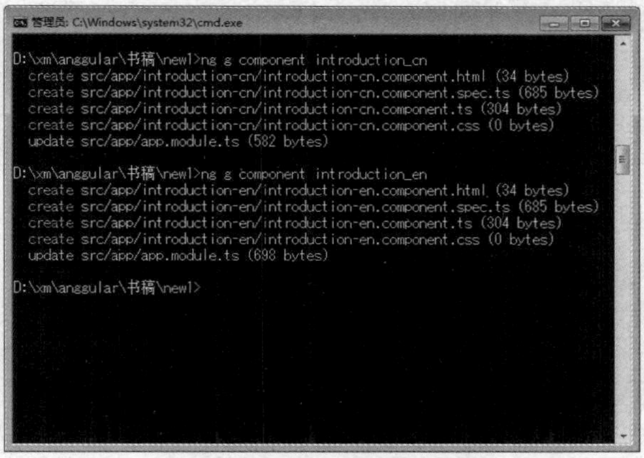

图 6.8　新建子组件

第二步：修改路由配置，在主路由中加上子路由（children 数组），当访问路径时加载对应的模板。代码如 CORE0616 所示。

代码 CORE0616：app-routing.module.ts

```
import {IntroductionEnComponent} from
        './introduction-en/introduction-en.component';
import {IntroductionCnComponent} from
        './introduction-cn/introduction-cn.component';

const routes : Routes = [
   { path: 'books/:bookname',component: BooksComponent,
        children: [
           {path: ' ', component: IntroductionCnComponent},
           {path: 'en', component: IntroductionEnComponent}
   ],
     },
];
```

第三步：在主路由模板上加子路由插座（router-outlet）。指定子路由组件显示的位置。代码如 CORE0617 所示。

代码 CORE0617：books.component.html

```
<div class="book">
    <p>
 页面二内容
    </p>
    <a [routerLink]="['./']"> 中文 </a>
    <a [routerLink]="['./en']"> 英文 </a>
    <router-outlet></router-outlet>
</div>
```

第四步：编写子路由组件。代码如下所示。

```
<p>
  在主路由的插座也就是出口处定义一个辅助路由插座
</p>
```

4 辅助路由

辅助路由是一个页面使用多个插座，并同时控制每个插座所显示内容的路由，相对于基本路由（组件模板上只有一个插座），辅助路由更加灵活。使用辅助路由步骤如下：

第一步：在主路由的插座定义一个辅助路由插座。添加一个 name 属性用来指定辅助路由显示的组件。代码如下所示。

```
<router-outlet></router-outlet>
<router-outlet name="xxx"></router-outlet>
```

第二步：配置入口参数。添加 outlets 属性（是一个对象，它的值也是一个对象），该对象里面传入上述的 name 属性值，指定要显示的辅助路由名字，值是该辅助路由需要显示的组件路径。代码如 CORE0618 所示。

代码 CORE0618：配置入口参数

```
// 名字叫 xxx 的辅助路由将显示路径为 'page1' 的组件
<a [routerLink]="[{outlets:{xxx: ' 辅助路由名字 '}}]"> 开始 </a>
// 当不希望辅助路由显示的时候可以把 name 设置为 null。
<a [routerLink]="[{outlets:{xxx:null}}]"> 结束 </a>
```

当希望跳转辅助路由的同时，主路由跳转到指定的组件，可以在入口文件添加 primary 属性：其属性值是需要跳转的主组件的路由路径。

```
<a [routerLink]="[{outlets:{primary:' 路由路径 ', advise:' xxx '}}]"> 开始 </a>
<a [routerLink]="[{outlets:{advise:null}}]"> 结束 </a>
```

第三步：配置辅助路由路径。添加 outlet 属性，指定该路由显示在对应名称的路由插座上。代码如 CORE0619 所示。

代码 CORE0619：app-routing.module.ts

```
{path: 'page1',component:Page1Component,outlet="xxx"}
```

使用辅助路由效果如图 6.9 所示。

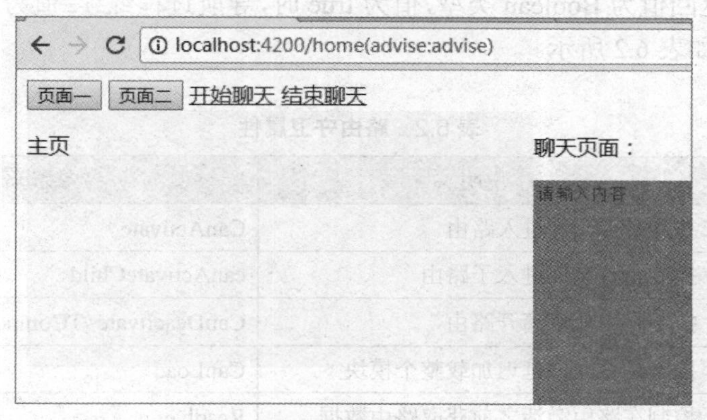

图 6.9　辅助路由效果

为了实现图 6.9 效果，代码如 CORE0620 所示。点击开始聊天的同时无论目前在哪个组件下主路由都会跳回 home 路径下的组件。

代码 CORE0620：app.component.html

```
<a [routerLink]="[{outlets:{primary:'home', advise:'advise'}}]"> 开始聊天 </a>
<a [routerLink]="[{outlets:{advise:null}}]"> 结束聊天 </a>
<router-outlet></router-outlet>
<router-outlet name="advise"></router-outlet>
```

配置路由，代码如 CORE0621 所示。

代码 CORE0621：app-routing.module.ts

```
import {IntroductionEnComponent} from './introduction-en/introduction-en.component';
import {IntroductionCnComponent} from './introduction-cn/introduction-cn.component';

const appRoutes: Routes = [
{path:'advise',component:AdviseComponent,outlet:"advise"},
];
```

聊天界面代码如 CORE0622 所示。

代码 CORE0622: advise.component.html

```
<p> 聊天页面:</p>
<textarea placeholder=" 请输入内容 " class="advise"></textarea>
```

5 路由守卫

Angular 路由守卫允许从一个页面跳转到另一个页面时执行一些指定的逻辑,并根据执行的结果来决定是否跳转。其实现效果为当用户满足一些条件后才允许进入或离开路由。当离开一个页面时,其返回值为 Boolean 类型,值为 true 时,导航过程继续;值为 false 时,导航被取消。路由守卫属性如表 6.2 所示。

表 6.2 路由守卫属性

属性	说明	接口名
CanActivate	控制是否允许进入路由	CanActivate
canActivateChild	控制是否允许进入子路由	canActivateChild
CanDeactivate	控制是否允许离开路由	CanDeactivate<TComponent>
canLoad	控制是否允许延迟加载整个模块	CanLoad
Resolve	控制在路由激活之前获取路由数据	Resolve

第一步:新建守卫服务文件(PermissionGuard.ts),添加导入 PermissionGuard 类(可变)并实现 Angular 框架提供的守卫接口,每个接口都需要实现相应的方法,就下而论,继承 CanActivate 并实现一个叫 canActivate 的方法,且返回一个布尔类型的值。代码如 CORE0623 所示。

代码 CORE0623: PermissionGuard.ts

```
import { Injectable }        from '@angular/core';
import { CanActivate, Router } from'@angular/router';

@Injectable()
//AuthGuard 类是需要继承 CanActivate 接口
export class PermissionGuard implementsCanActivate {
 constructor(private authService: AuthService, private router: Router) {}

 canActivate() {
// 一些逻辑判断代码
   }
}
```

第二步：修改路由配置，把守卫属性加到路由中。代码如 CORE0624 所示。

代码 CORE0624：app-routing.module.ts

```
import { PermissionGuard   } from './../PermissionGuard';

const appRoutes: Routes = [
{ path: 'admin', component: GuardAdminComponent, canActivate: [PermissionGuard] }
];
```

第三步：在 app.module.ts 文件中导入守卫路由的服务。代码如下所示。

```
import {PermissionGuard} from './guard/PermissionGuard';

@NgModule({
  providers: [PermissionGuard],
})
export class AppModule {
}
```

使用路由守卫实现以下效果，当点击链接书籍时会先判断生成的随机数除 2 时的余数，余 0 则可进入，否则被阻止。效果如图 6.10 所示。

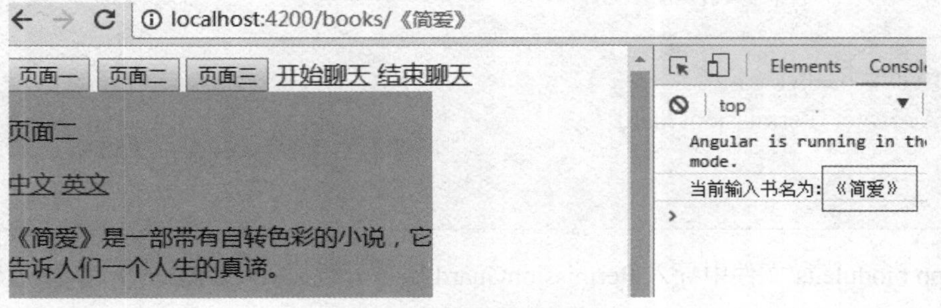

图 6.10　路由守卫效果

为了实现图 6.10 效果，代码如 CORE0625 所示。

代码 CORE0625：app.component.html

```
<input type="button" value=" 页面三 " (click)="toBookDetails()">
```

新建 PermissionGuard.ts 文件，为进入页面三之前加入逻辑判断，点击页面三时会先判断生成的随机数除 2 时的余数，余 0 则可进入，否则被阻止。代码如 CORE0626 所示。

代码 CORE0626：PermissionGuard.ts

```
// 导入 CanActivate 参数
```

```
import {ActivatedRouteSnapshot, CanActivate, RouterStateSnapshot}
from '@angular/router';
import {Observable} from 'rxjs/Observable';
import * as _ from 'lodash';

export class PermissionGuard implements CanActivate {
// 再次点击页面三时会先判断生成的随机数除 2 时的余数，余 0 则可进入，否则被
// 阻止。
  canActivate(route: ActivatedRouteSnapshot,
              state: RouterStateSnapshot): boolean | Observable<boolean> | Promise<bool-
ean> {
    const rand = _.random(0, 6);
    console.log(rand);
    return rand % 2 === 0;
  }
}
```

配置路由，增加路由守卫。代码如 CORE0627 所示。

代码 CORE0627：app-routing.module.ts

```
import {PermissionGuard} from './guard/PermissionGuard';

const routes: Routes = [
  canActivate: [PermissionGuard]
  },
];
```

在 app.module.ts 文件中导入 PermissionGuard 服务。代码如 CORE0628 所示。

代码 CORE0628：app. module.ts

```
import {BrowserModule} from '@angular/platform-browser';
import {NgModule} from '@angular/core';

import {AppRoutingModule} from './app-routing.module';
import {AppComponent} from './app.component';
import {HomeComponent} from './home/home.component';
import {BooksComponent} from './books/books.component';
import {IntroductionEnComponent} from './introduction-en/introduction-en.component';
import {IntroductionCnComponent} from './introduction-cn/introduction-cn.component';
```

```
import {PermissionGuard} from './guard/PermissionGuard';
@NgModule({
  declarations: [
    AppComponent,
    HomeComponent,
    BooksComponent,
    IntroductionEnComponent,
    IntroductionCnComponent,
  ],
  imports: [
    BrowserModule,
    AppRoutingModule
  ],
  providers: [PermissionGuard],
  bootstrap: [AppComponent]
})
export class AppModule {
}
```

在 app.component.ts 文件中引入 Router 模块并编写跳转的方法。代码如 CORE0629 所示。

代码 CORE0629：app.component.ts

```
import {Component} from '@angular/core';
import {Router} from '@angular/router';

@Component({
  selector: 'app-root',
  templateUrl: './app.component.html',
  styleUrls: ['./app.component.css']
})
export class AppComponent {
  constructor(private router: Router) {
  }
  toBookDetails() {
    this.router.navigate(['/books', '《简爱》']);
  }
}
```

通过下面八个步骤的操作，实现图 6.2 所示的智慧工厂气报表模块的效果。

第一步：在命令窗口输入以下命令创建路由文件、气报表、气报表详情、存储数据的服务组件。命令如下所示。

```
ng g module gas-reporting
ng g component gas-reporting/gas-detail
ng g component gas-reporting/gas-reporting
ng g service gas-reporting/product
```

第二步：在 gas-reporting.module.ts 文件中注入 ProductService 服务并引入路由文件。代码如 CORE0630 所示。

代码 CORE0630：gas-reporting.module.ts

```
import { NgModule } from '@angular/core';
import { CommonModule } from '@angular/common';
import { GasReportingComponent } from './gas-reporting/gas-reporting.component';
import { GasDetailComponent } from './gas-detail/gas-detail.component';
import {ProductService} from "./product.service";
import { Routes } from "@angular/router";
import { RouterModule } from '@angular/router';

@NgModule({
  imports: [
    CommonModule,
    RouterModule
  ],
  providers: [ProductService],
  declarations: [GasReportingComponent, GasDetailComponent,]
})
export class GasReportingModule { }
```

第三步：在 gas-reporting.component.html 文件中对组件进行布局，设置各个组件的渲染位置。代码如 CORE0631 所示。

代码 CORE0631：gas-reporting.component. html

```
<div style="background-color: white;height: 900px">
```

```
<h2 class="text-center"> 智慧工厂气报表信息查询 </h2>
<br>
  <div class="row" style="margin-left: 60px">
    <div class="col-md-11">
    <!-- 气报表表格 -->
    </div>
  </div>
<div class="row" style="margin-left: 60px">
  <div class="col-md-11">
  <!-- 底部空气质量指数组件 -->
  </div>
</div>
</div>
```

第四步：在 product.service.ts 服务文件定义一个类并填充数据，定义获取所有数据的方法，并根据数据 id 在数组找到对应数据。代码如 CORE0632 所示。

代码 CORE0632：表格 ts 代码

```
import { Injectable } from '@angular/core';

@Injectable()
export class ProductService {
  private products:Product[]=[
    new Product(1,"2017-09-01"," 锅炉大气污染物排放标准 ", "44","87"," 二氧化硫 ","Boston,  MA"," 不适合外出散步 ", "31 摄氏度 "),
    new Product(2,"2017-09-01"," 火电厂大气污染物排放标准 ", "66","80"," 颗粒物 ","Boston, MA"," 不适合外出散步 ", "31 摄氏度 "),

// 部分代码省略
  ];
  constructor() { }
// 获取所有商品的方法 , 返回商品数组
  gerProducts():Product[]{
    return this.products;
  }
  // 根据商品 id 在数组找到对应商品
  getProduct(id:number):Product{
    return this.products.find((product)=>product.id==id);
```

```
  }
}

// 定义一个类
export class Product{
  constructor(
    public id:number,
    public StartTime:string,
    public EndTime:string,
    public title:string,
    public reports:string,
    public city:string,
    public play:string,
    public temp:string
  ){
  }
}
```

在 gas-reporting.component.html 中，设置气报表，表格数据通过使用 NgFor 指令遍历循环显示。代码如 CORE0633 所示。

代码 CORE0633：左侧导航栏

```
<table class="table  table-striped table-bordered">
  <thead>
  <tr>
    <th> 编号 </th>
    <th> 报警时间 </th>
    <th> 执行标准名称 </th>
    <th> 实测浓度 (mg/m3)</th>
    <th> 折算浓度 (mg/m3)</th>
    <th> 监测项目名称 </th>
    <th> 报警详情 </th>
  </tr>
  </thead>
  <tbody  *ngFor="let product of products" class="pub_a">
  <tr>
    <td>{{product.id}}</td>
    <td class="success">{{product.StartTime}}</td>
```

```
        <td class="warning">{{product.EndTime}}</td>
        <td class="danger">{{product.title}}</td>
        <td class="active">{{product.reports}}</td>
        <td class="success">{{product.reports1}}</td>
        <td ><a> 查看详情 </a></td>
    <tr>
    </tbody>
</table>
```

在对应界面的 ts 文件注入服务并声明一个变量用于存放数组，使用 gerProducts() 方法从服务中获取数据。代码如 CORE0634 所示。设置后效果如图 6.11 所示。

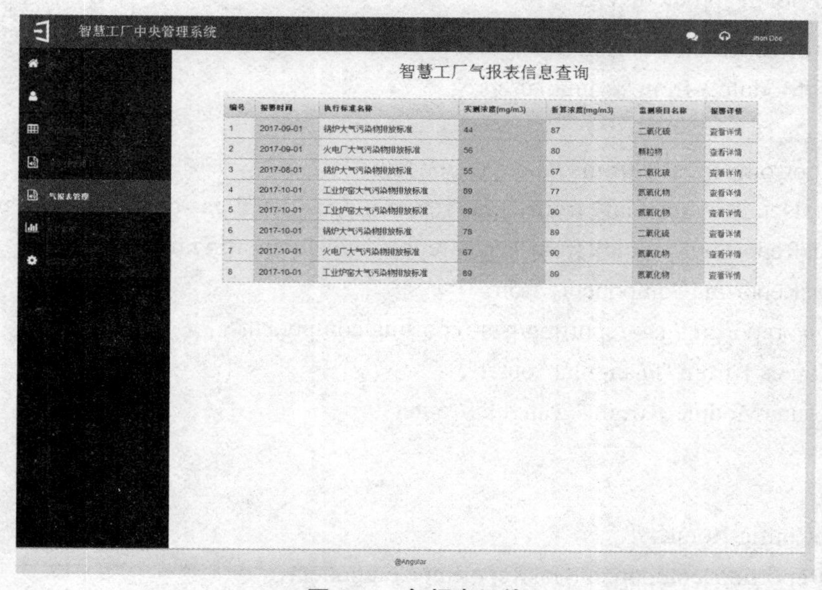

图 6.11　气报表组件

代码 CORE0634：gas-reporting.component.ts

```
import { Component, OnInit } from '@angular/core';


import {ProductService} from "../product.service";
import {Product} from "../product.service";
@Component({
  selector: 'app-gas-reporting',
  templateUrl: './gas-reporting.component.html',
  styleUrls: ['./gas-reporting.component.css']
})
```

```
export class GasReportingComponent implements OnInit {
  // 定义一个数组
  public products:Product[];
  constructor(private productService:ProductService) { }
  ngOnInit() {
    this.products=this.productService.gerProducts();
  }
}
```

第五步：在 app.module.ts 文件配置路由，代码如 CORE0635 所示。

代码 CORE0635：app.module.ts

```
import { BrowserModule } from '@angular/platform-browser';
import { NgModule } from '@angular/core';

import { AppComponent } from './app.component';
import {GasDetailComponent} from "./gas-reporting/gas-detail/gas-detail.component";
import {GasReportingModule} from "./gas-reporting/gas-reporting.module";
import {GasReportingComponent} from
        "./gas-reporting/gas-reporting/gas-reporting.component";
import { Routes } from "@angular/router";
import { RouterModule } from "@angular/router";

// 路由配置
const routeConfig:Routes=[
{path:'gasReporting',component:GasReportingComponent},
{path:'product/:productId',component:GasDetailComponent},
]
@NgModule({
  declarations: [
  AppComponent,
    ],
  imports: [
  GasReportingModule,
  BrowserModule,
  BrowserAnimationsModule,
```

```
  ],
  providers: [],
  bootstrap: [AppComponent]
})
export class AppModule { }
```

第六步：在 gas-reporting.component.html 文件中修改气报表组件，给报表添加 routeLink 指令。当点击查看详情时，通过 id 把数据传入详情页。代码如 CORE0636 所示。

代码 CORE0636：气报表 HTML 代码

```
<td ><a [routerLink]="['/product',product.id]"> 查看详情 </a></td>
```

第七步：在 gas-detail.component.ts 文件中，注入保存当前信息的对象（ActivatedRoute），通过参数快照，在导航到气报表详情页时传递气报表的 id。代码如 CORE0637 所示。

代码 CORE0637：表格 ts 代码

```
import { Component, OnInit } from '@angular/core';
import { ActivatedRoute } from '@angular/router';
import {Product} from "../product.service";
import {ProductService} from "../product.service";
@Component({
  selector: 'app-gas-detail',
  templateUrl: './gas-detail.component.html',
  styleUrls: ['./gas-detail.component.css']
})
export class GasDetailComponent implements OnInit {
  product:Product;
// 注入服务并使用 ActivatedRoute 对象保存获取的数据
  constructor(private  routeIofo:ActivatedRoute,
              private  productService:ProductService) {}

  ngOnInit() {
// 通过 id 获取数据
    let productId:number=this.routeIofo.snapshot.params["productId"]
    this.product=this.productService.getProduct(productId);
  }
}
```

在 product.service.ts 文件中写入根据数据 id 在数组找到对应数据的方法。

```
import { Injectable } from '@angular/core';

@Injectable()
export class ProductService {
  // 根据数据 id 在数组找到对应数据
  getProduct(id:number):Product{
    return this.products.find((product)=>product.id==id);
  }
}
```

在 gas-detail.component.html 中，设置气报表详情界面，通过 <h2> 标签设置表格标题，数据通过表达式显示，数据变量名需要和气报表表格一致，以保证数据可以通过 id 传入到详情页。代码如 CORE0638 所示。设置后效果如图 6.12 所示。

代码 CORE0638：gas-detail.component.html

```
<div style="height: 860px">
<h2 class="text-center"> 智慧工厂气报表详细信息 </h2>
  <div class="row" style="margin-left: 60px">
    <div class="col-md-11">
<table class="table table-striped table-bordered" style="margin-top: 10px">
  <tr>
    <th> 编号 </th>
    <td>{{product.id}}</td>
  </tr>
  <tr>
    <th> 报警时间 </th>
    <td>{{product.StartTime}}</td>
  </tr>
  <tr>
    <th> 执行标准名称 </th>
    <td>{{product.EndTime}}</td>
  </tr>
  <tr>
    <th> 实测浓度 (mg/m3)</th>
    <td>{{product.title}}</td>
  </tr>
  <tr>
    <th> 折算浓度 (mg/m3)</th>
```

```
            <td>{{product.reports}}</td>
        </tr>
        <tr>
            <th> 监测项目名称 </th>
            <td>{{product.city}}</td>
        </tr>
        <tr>
            <th> 室外温度 </th>
            <td>{{product.temp}}</td>
        </tr>
</table>
    <button [routerLink]="['/gasReporting']"
    style="float: right;margin-right: 20px">
    <p style="margin: 0px"> 返回 </p></button>
</div>
    </div>
</div>
```

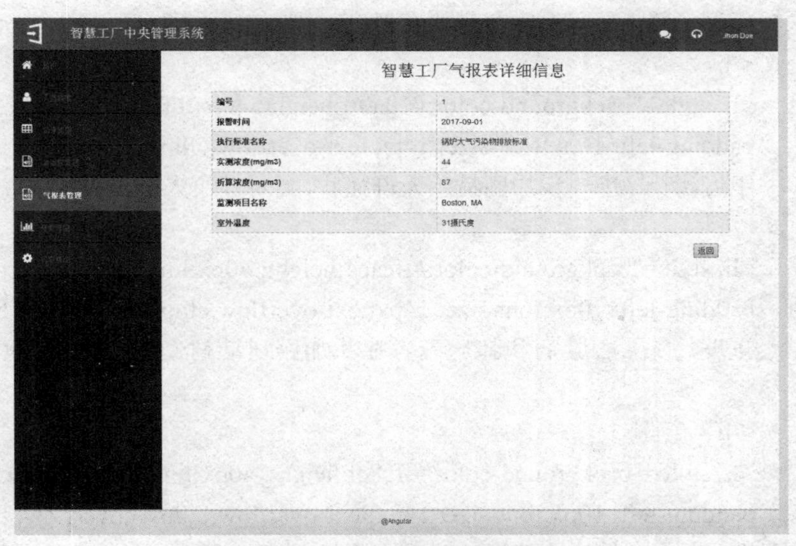

图 6.12　气报表详细信息组件

第八步：制作质量指数组件。在 gas-reporting.component.html 中，通过栅格系统布局、无序列表显示气信息小贴士的内容。代码如 CORE0639 所示。设置样式前效果如图 6.13 所示。

代码 CORE0639：HTML 代码

```
<div class="row" style="margin-left: 60px">
    <div class="col-md-11">
```

```
    <div class="widget green">
        <div class="widget-title">
            <h4 style="color: white"><i class="icon-reorder"></i> 气信息显示小贴士 </h4>
        </div>
        <div class="widget-content" style="padding-top: 10px;padding-right: 20px">
            <div class=" ">
                <div class="clearfix"></div>
                <ol id="slist">
                    <li  style="background-color:#efffff;height: 40px;line-height: 40px;
                    padding-left: 10px;font-size: 14px;text-overflow:ellipsis;overflow: hidden">
                    优：空气质量令人满意，基本无空气污染。
                    </li>
                    <li  style="background-color:#47990a;height: 40px;line-height: 40px;
                    padding-left: 10px;font-size: 14px;text-overflow:ellipsis;overflow: hidden">
                    良：空气质量可接受，某些污染物对极少数敏感人群健康有较弱影响。
                    </li>
                    <li  style="background-color:#06c4ff;height: 40px;line-height: 40px;
                    padding-left: 10px;font-size: 14px;text-overflow:ellipsis;overflow: hidden">
                    轻度污染：易感人群有症状有轻度加剧，健康人群出现刺激症状。
                    </li>
                    <li  style="background-color:#10a0ff;height: 40px;line-height: 40px;
                    padding-left: 10px;font-size: 14px;text-overflow:ellipsis;overflow: hidden">
                    中度污染：进一步加剧易感人群症状，会对健康人群的呼吸系统有影响。
                    </li>
                    <li  style="background-color:#ffeac4;height: 40px;line-height: 40px;
                    padding-left: 10px;font-size: 14px;text-overflow:ellipsis;overflow: hidden">
                    重度污染：心脏病和肺病患者症状加剧运动耐受力降低，健康人群出现
症状
                    </li>
                    <li  style="background-color:#ffc8cf;height: 40px;line-height: 40px;
                    padding-left: 10px;font-size: 14px;text-overflow:ellipsis;overflow: hidden">
                    严重污染：健康人群运动耐受力降低，有明显强烈症状，可能导致疾病
                    </li>
                </ol>
            </div>
        </div>
    </div>
</div>
```

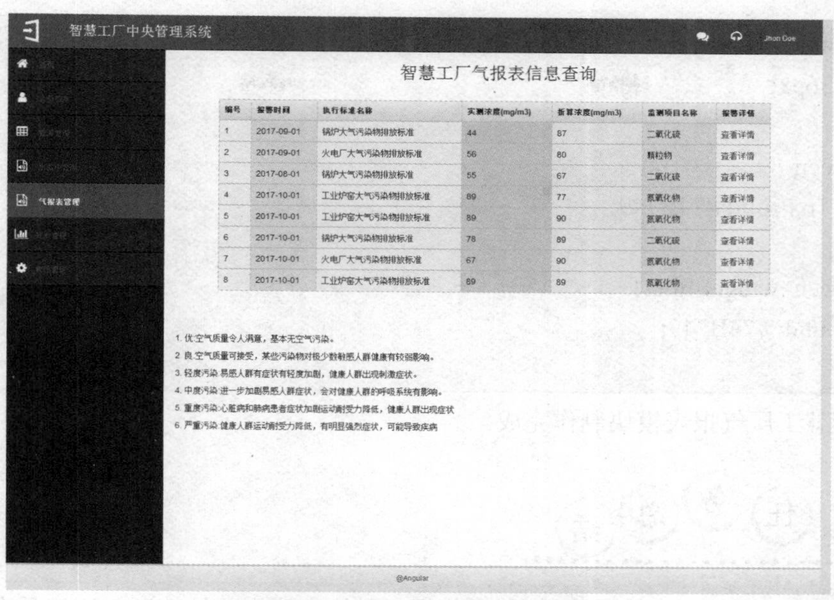

图 6.13　质量指数组件设置效果前

通过 CSS 样式设置质量指数组件中的文字大小、间距等，给整体内容增加边框并设置字体背景颜色，使其更加美观。

```
.widget-title > h4 {
  float: left;
  font-size: 14px;
  font-weight: normal;
  padding: 10px 11px 10px 16px;
  line-height: 12px;
  margin: 0;
}
.widget-title > h4 i {
  font-size: 14px;
  margin-right: 2px;
}
.widget {
  background:#fff;
  clear: both;
  margin-bottom: 20px;
  margin-top: 0;
  border: 1px solid #404040;
}
.widget-title {
```

```
  background: #404040;
  height: 36px;
}
.widget.green {
  border: 1px solid #74B749;
}
.widget.green .widget-title {
  background: #74B749;
}
```

至此，智慧工厂气报表模块制作完成。

通过对智慧工厂气报表模块的学习，对气报表模块中路由跳转等所需知识具有初步了解，掌握气报表模块的页面跳转的流程，了解 Angular 路由的使用方法，具有独立运用 Angular 路由基本用法的能力，为界面导航打下一定基础。

route　路由
redirectTo　重定向
prefix　前缀
params　参数
guards　守卫
resolve　解决
ActivedRoute　激活的路由
routerLink　路由链接
outlet　出口

一、选择题

1. 以下哪个（　）在配置路由，可以保存 URL 对应的组件。

（A）Routes　　（B）RouterOutlet　　（C）Router　　（D）routerLink

2. 以下哪个（　）表示当前激活的路由对象，保存当前路由的信息，如路由地址、参数。

(A)RouterModule　(B)ActivedRoute　(C)redirectTo　(D)routerLinkActive

3. 在主路由的插座定义一个辅助路由插座。添加一个(　　)属性用来指定辅助路由显示的组件。

(A)Routes　(B)RouterModule　(C)name　(D)Router

4. 配置辅助路由路径。添加(　　)属性,指定该路由显示在对应名字的辅助路由插座上。

(A)outlet　(B)name　(C)scope　(D)bind

5. 路由传参在详情页使用(　　)对象来接收参数。

(A)ActivatedRoute　(B)outlet　(C)CanActivate　(D)CanDeactivate

二、填空题

1.________ 是在 HTML 中声明路由导航的指令。

2. 使用 angular-cli 创建一个带路由的项目格式为 ________。

3. 使用路由需要在 AppModule 模块中导入 ________。

4. RouterModule.forRoot() 方法用于在主模块中定义路由信息,通过调用该方法使主模块可以访问路由模块中定义的所有指令,与 Router.forRoot() 方法类似,但它只能应用在特性模块中,子模块中使用 ________。

5. ________ 路由一般放在最后面,如果当前 URL 无法匹配上配置过的任何一个路由中的路径。

三、上机题

使用 Angular 编写符合以下要求的页面。要求:

1. 头部为选项卡,内容部分采用列表布局,当点击其中一个列时跳转到相应详情页,并把对应的参数传过去。

2. 使用 Angular 路由、路由传参,效果如下图。

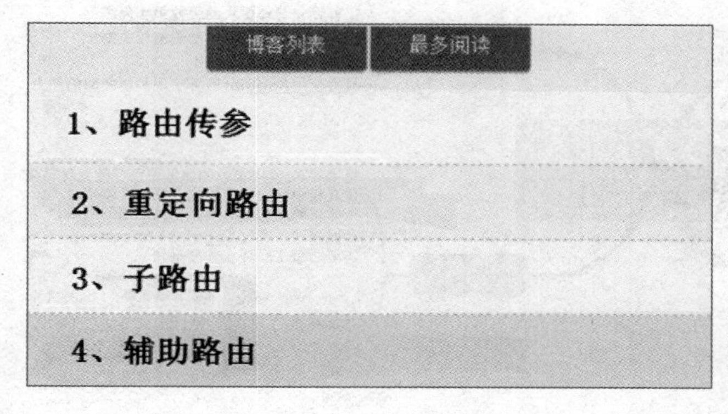

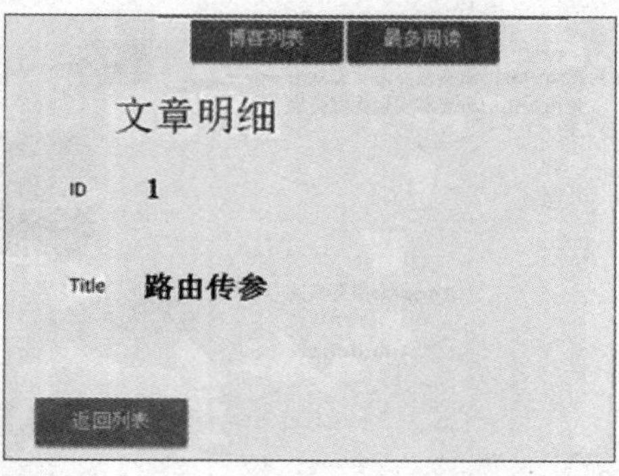

项目七　智慧工厂环安管理模块

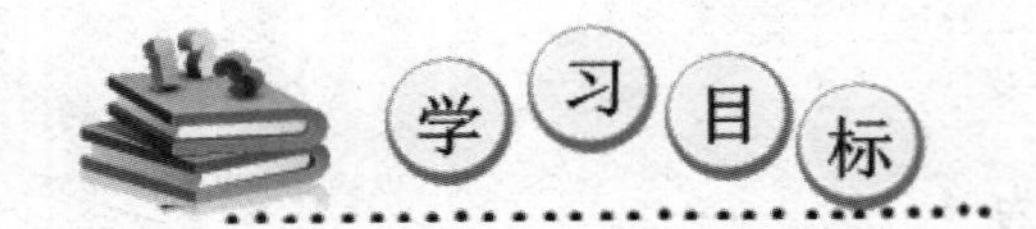

通过智慧工厂环安管理模块功能的实现，了解环安管理模块的动画效果实现流程，学习Angular 服务、动画的相关知识，掌握 HTTP 服务的使用，具有使用 Angular 服务实现数据传输的能力。在任务实现过程中：

- 了解环安管理模块动画效果实现流程。
- 学习 Angular 服务、动画的相关知识。
- 掌握 HTTP 服务的使用。
- 具有使用 Angular 服务实现数据传输的能力。

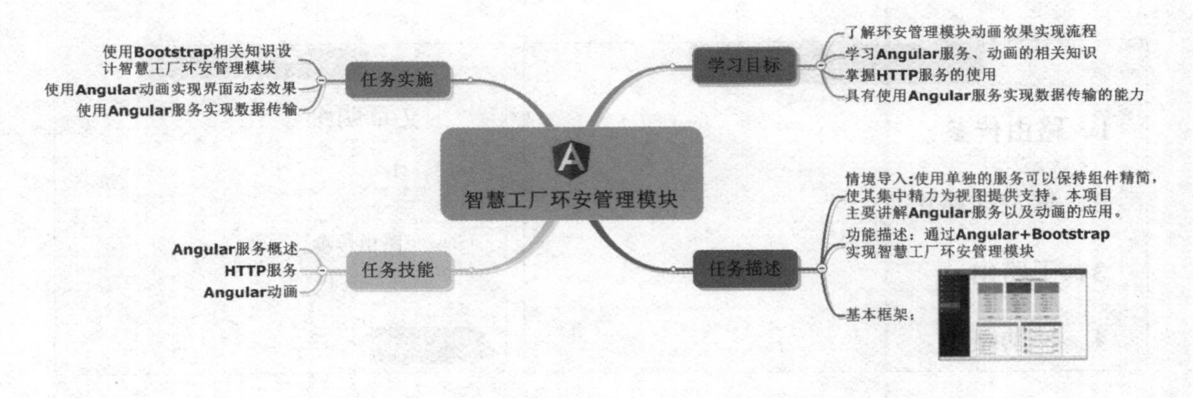

【情境导入】

在智慧工厂中央管理系统的制作过程中，路由管理和视图加载具有一定的重要性，当两个视图数据一致时，我们也许会把代码一遍又一遍的复制，智慧工厂的开发人员觉得这样会很麻烦，所以他们使用服务提供一种能在应用的整个生命周期内保持和共享数据的方法，当用到时只需把它注入到相应的组件中。使用服务可以保持组件精简，使其集中精力为视图提供支持。本项目主要讲解 Angular 服务以及动画的应用。

【功能描述】

使用 Bootstrap+Angular 实现智慧工厂环安管理模块：

- 使用 Bootstrap 相关知识设计智慧工厂环安管理模块。
- 使用 Angular 动画实现界面动态效果。
- 使用 Angular 服务实现数据传输。

【基本框架】

基本框架如图 7.1 所示，通过本项目的学习，能将图 7.1 所示的框架图转换成智慧工厂环安管理模块并实现其功能，效果图如图 7.2 所示。

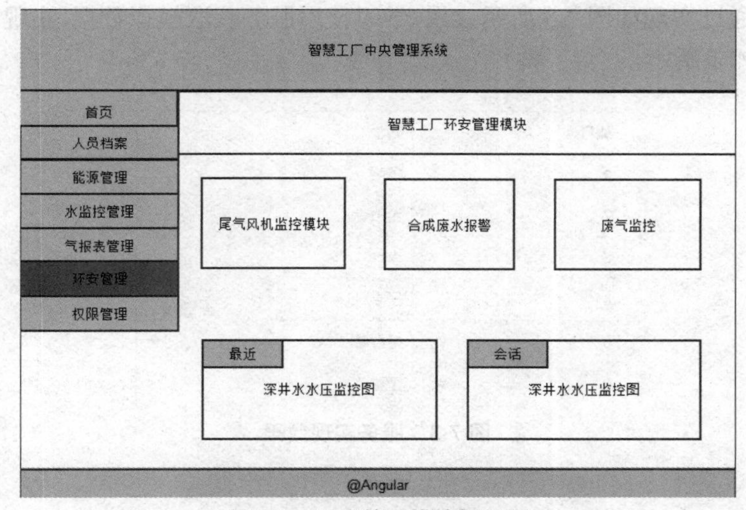

图 7.1　环安管理模块框架图

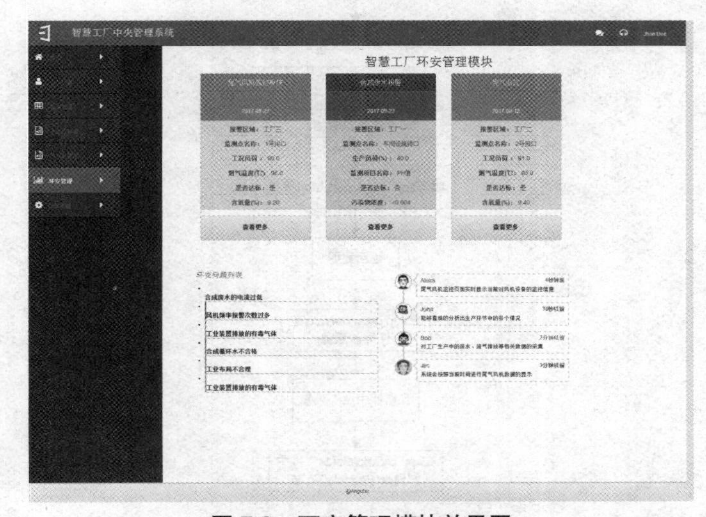

图 7.2　环安管理模块效果图

技能点 1 Angular 服务概述

Angular 服务是指能够被其他的组件或者指令调用的、可共享的代码块，当有组件需要时，通过依赖注入，将其注入到组件中。具有提高代码的利用率，方便组件之间共享数据和方法，方便测试和维护等优点。Angular 服务使用非常广泛，例如，当需要相同数据的组件时，把需要相同的数据存储在服务中，在需要的组件注入服务实现数据共享。

下面，我们通过实现如图 7.3 所示效果，学习使用服务获取模拟数据，实现流程图如图 7.4 所示，具体步骤如下所示。

编号	名字
1	杨明
2	金浩森
3	李烨
4	王宇
5	姬存希
6	王雅

图 7.3 服务实现效果

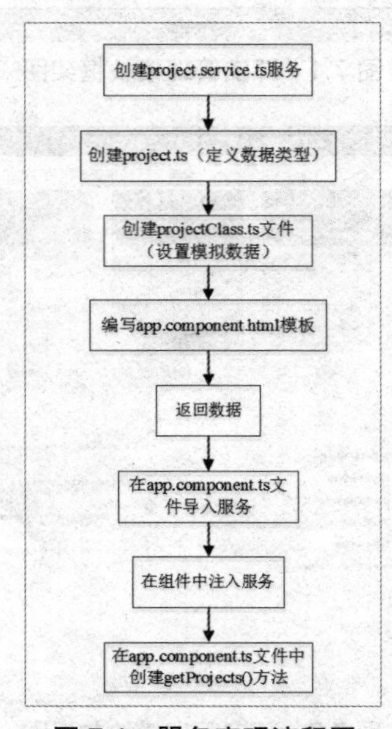

图 7.4 服务实现流程图

第一步：创建服务。

打开命令窗口，使用 ng g service project（服务名称）命令，在 app 目录下创建 project.service.ts 服务文件。从 @angular/core 中导出了 Injectable（告诉其他引用的组件这个服务是可注入的），代码如下所示。

```
import { Injectable } from '@angular/core';
  @Injectable()
export class ProjectService {
}
```

第二步：声明变量。

为了使数据处理更加方便，需要声明变量，创建 project.ts 文件，为其定义 id 与 name 的数据类型，代码如下所示。

```
export class Project {
  id: number;
  name: string;
}
```

第三步：模拟数据。

创建 projectClass.ts 文件，设置模拟数据。然后在该文件中使用 import 导入 Project 类。注：模拟数据的数据类型要与 Project 类中定义的数据类型相匹配，代码如下所示。

```
import { Project } from './project';
export const PROJECTS: Project[] = [
  { id: 1, name: ' 杨明 ' },
  { id: 2, name: ' 金浩森 ' },
  { id: 3, name: ' 李烨 ' },
  { id: 4, name: ' 王宇 ' },
  { id:5, name: ' 姬存希 ' },
  { id:6, name: ' 王雅 ' },
];
```

第四步：编写 app.component.html 模板文件显示数据，代码如下所示。

```
<table class="table table-striped table-bordered">
  <tr>
    <th> 编号 </th>
    <th> 名字 </th>
  </tr>
  <tr  *ngFor="let project of projects">
```

```
    <td>{{project.id}}</td>
    <td>{{project.name}}</td>
  </tr>
</table>
```

第五步：返回数据。

在 ProjectService 服务中，导入 PROJECTS 常量及 Project 类，并在 getProducts() 方法（可变）中返回数据，代码如下所示。

```
import { Injectable } from '@angular/core';
import { Project } from './project';
import { PROJECTS } from './projectclass';
@Injectable()
export class ProjectService {
// 在 getProducts() 方法（可变）中返回数据
  getProjects(): Promise<Project[]> {
    return Promise.resolve(PROJECTS);
  }
}
```

第六步：导入服务。

在 app.component.ts 文件（需要服务的组件）中通过 import 导入 ProjectService 服务，代码如下所示。

```
import { ProjectService } from './project.service';
```

接下来，在 app.component.ts 文件中，添加尚未初始化的 projects 属性，代码如下所示。

```
projects: Project[];
```

第七步：依赖注入服务。

为了在组件中获取到 ProjectService 服务，需要依赖注入所需服务，在这个例子中，采用在组件中注入服务（项目五中具有在组件中依赖注入的讲解），代码如下所示。

```
import { Project } from './project';
import { ProjectService } from './project.service';
@Component({
  selector: 'my-app',
  templateUrl: './app.component.html',
  providers: [ProjectService]
})
```

```
export class AppComponent implements OnInit {
  projects: Project[];
  constructor(private projectService: ProjectService) { }
}
```

第八步：获取数据。

在 app.component.ts 文件中创建 getProjects() 方法（可更改），并通过该方法获取数据，代码如下所示。

```
import { Component, OnInit } from '@angular/core';
import { Project } from './project';
// 导入服务
import { ProjectService } from './project.service';
@Component({
  selector: 'my-app',
  templateUrl: './app.component.html',
// 注入服务
  providers: [ProjectService]
})
export class AppComponent implements OnInit {
  projects: Project[];
    // 定义私有属性
  constructor(private projectService: ProjectService) { }
    // 获取数据
  getProjects(): void {
    this.projectService.getProjects().then(projects => this.projects = projects);
  }
  ngOnInit(): void {
    this.getProjects();
  }
}
```

提示:无论在生活还是在工作中,只有积极的人才能在每一次忧患中看到希望,而消极的人则看不到一丝希望。你是一个积极的人吗?是一个享受生活的人吗?快来扫我吧!你将会收获快乐。

技能点 2　HTTP 服务

大多数前端应用都需要通过 HTTP 协议与后端服务器进行通讯。在前端开发中,一般通过原生 Ajax 进行服务器的访问,但在访问服务器时会发生跨域问题,通过 JSONP 可以解决这一问题。本项目主要通过 Angular 封装 Ajax 后的 HTTP 服务进行讲解。

1　Ajax 介绍

Ajax 是使用 XMLHttpRequest 对象(支持同步和异步的方式发送请求,默认用异步方式)向服务器请求并操作数据的一种通讯传输技术。通过 XMLHttpRequest 对象来向服务器发送请求,从服务器获得数据,然后用 JavaScript 来操作 DOM 更新页面,具有无需刷新即可更新数据的优点。

XMLHttpRequest 是 Ajax 的核心,主要作用是在客户端和服务器之间传输数据,通过 URL 即可获取数据,且不会刷新整个页面(局部刷新)。它具有很多属性和方法。部分属性如表 7.1 所示。

表 7.1　部分属性表

属性	描述
readyState	unsigned short—请求的状态
response	varies—响应体的类型由 responseType 来指定,可以是 ArrayBuffer、Blob、Document、JSON 或字符串,如果请求未完成或失败,则该值为 null
responseText	DOMString—此请求的响应为文本等
responseType	XMLHttpRequestResponseType—设置该值能够改变响应类型
responseXML	Document—本次请求响应是一个 Document 对象
status	unsigned short—请求的响应状态码

续表

属性	描述
statusText	DOMString—请求的响应状态信息，包含一个状态码和消息文本
onreadystatechange	Function—当 readyState 属性改变时会被调用

其中 readyState 具有五种表示请求的状态：

- UNSENT（未打开）：表示已创建 XHR 对象，open() 方法还未被调用。
- OPENED（未发送）：open() 方法已被成功调用，send() 方法还未被调用。
- HEADERS_RECEIVED（已获取响应头）：send() 方法已经被调用，响应头和响应状态已经返回。
- LOADING（正在下载响应体）：响应体下载中，responseText 中已经获取了部分数据。
- DONE（请求完成）：整个请求过程已经完毕。

XMLHttpRequest 的方法如表 7.2 所示。

表 7.2　部分方法表

方法	描述
abort()	如果请求已经被发送，则立刻中止请求
getAllResponseHeaders()	返回所有响应头信息（响应头名和值），如果响应头还没有接收，则返回 null
getResponseHeader()	返回指定响应头的值，如果响应头还没有被接收，或该响应头不存在，则返回 null
open()	初始化一个请求
overrideMimeType()	重写由服务器返回的 MIME 类型
send()	发送请求。如果该请求是异步模式（默认），该方法会立刻返回。相反，如果请求是同步模式，则直到请求的响应完全接受以后，该方法才会返回
setRequestHeader()	设置 HTTP 请求头信息

HTTP 状态码：每发出一个 HTTP 请求之后，就会有一个响应，HTTP 本身会有一个状态码，来标示这个请求是否成功，常见状态码：

- 200，2 开头的都表示这个请求发送成功，最常见的就是 200。
- 300，3 开头的代表重定向，最常见的是 302。
- 400，400 代表客户端发送的请求有语法错误，401 代表访问的页面没有授权，403 代表没有权限访问这个页面，404 代表没有这个页面。
- 500，5 开头的代表服务器有异常，500 代表服务器内部异常，504 代表服务器端超时，没返回结果。

在 Angular 中使用 Ajax 处理数据，效果如图 7.5 所示。

ID	登录名
4254276	alexpods
1446119	asnowwolf
757236	bradlygreen
3429878	bradrich
782920	calebegg

图 7.5 Ajax 处理数据

为了实现图 7.5 所示效果，在 app.component.html 文件中，创建 HTML 模板，代码如 CORE0701 所示。

代码 CORE0701：HTML 模板

```
<table>
  <tr>
    <th>ID</th>
    <th> 登录名 </th>
  </tr>
  <tr *ngFor="let member of members;">
    <td>{{member.id}}</td>
    <td>{{member.login}}</td>
  </tr>
</table>
```

对应的 ts 代码如 CORE0702 所示。

代码 CORE0702：app.component.ts 文件

```
import { Component, OnInit } from '@angular/core';
interface Member {
  id: number;
  login: string;
}
@Component({
  selector: 'app-root',
  templateUrl: './app.component.html',
  styleUrls: ['./app.component.css']
})
export class AppComponent implements OnInit {
  members: Member[];
  getMembers() {
  // 路径为网上找的
```

```
    let MEMBERS_URL = `https://api.github.com/orgs/angular/members?page=1&per_page=5`;
    // 创建 XMLHttpRequest 对象
    let xhr = new XMLHttpRequest();
    // 设置请求方式和请求 URL 地址
    xhr.open("GET", MEMBERS_URL);
    // 监听 readyState 状态变化
    xhr.onreadystatechange = () => {
      // 判断请求是否完成且请求成功
      if (xhr.readyState == 4 && xhr.status == 200) {
        if (xhr.responseText) {
          try {
            // 把响应体转换为 JSON 对象，并赋值给 members 属性
            this.members = JSON.parse(xhr.responseText);
          } catch (error) {
            throw error;
          }
        }
      }
    };
    // 发送 HTTP 请求
    xhr.send(null);
  }
  ngOnInit() {
    this.getMembers();
  }
}
```

2　JSONP 介绍

JSONP 是用于解决主流浏览器的跨域访问数据的问题。在 Web 开发中，有时需要向与当前页面不同源（域名、协议、端口不相同）的服务器发送 Ajax 请求，由于 XMLHttpRequest 只允许请求当前源的内容，所以该请求会被浏览器阻止，这时就需要用到 JSONP。其实现方式在客户端声明回调函数，通过 GET 方法向服务器跨域请求数据，然后服务端返回相应的数据并动态执行回调函数。

在 Angular 中包含了 JSONP 服务，使用 JSONP 服务，首先在构造函数中注入 JSONP 服务，然后定义 URLSearchParams 对象（用于创建 URL 参数）来构造请求参数，最后调用 JSONP 服务的 get() 方法（可变）发起请求。要想在服务类中运用，还需在根模块加载 JsonpModule。使用 JSONP 服务示例代码如下所示。

```
import { Injectable } from '@angular/core';
// 导入 Jsonp
import { Jsonp, URLSearchParams } from '@angular/http';
import 'rxjs/add/operator/map';
@Injectable()
export class AppService {
  // 注入 JSONP 服务，并构造请求参数
  constructor(private jsonp: Jsonp) {}
  // 发起请求
  getApps() {
  // 定义服务器地址
  let URL = ' 访问路径 ';
  // 定义参数
  let params = new URLSearchParams();
  params.set('foramt','json');
  params.set('callback','JSONP_CALLBACK');
  // 使用 map().substribe() 获取数据
  return this.jsonp
     .get(URL,{search: params})
     .map(response => response.json().app)
     .subscribe(
        apps => this.dates = apps,
        error => this.erroMsg = <any>error)
  }
}
```

注：JSONP 只能发起 GET 请求。

3 HttpModule

相对于使用 XMLHttpRequest API 实现图 7.3 效果，使用 HttpModule 也可以实现同样的功能，而且代码更加简洁。HttpModule 底层实现是基于 XMLHttpRequest API，并对 XMLHttpRequest API 进行了封装，抽象出了 Body、Request、Headers 和 Response 等对象。Request 类和 Response 类是继承于 Body 类，Body 类中提供了四个方法用于数据转换，具体如表 7.3 所示。

表 7.3 数据转换方法

方法	描述
json()	转换为 JSON 对象
text()	转换为 text 对象

续表

方法	描述
arrayBuffer()	转换为 ArrayBuffer 对象
blob()	转化为 Blob 对象

HttpModule 是用来封装 HTTP 服务相关功能的模块，该模块不仅包含了 HTTP 服务，而且也包含了 HTTP 所依赖的其他服务。HttpModule 主要包含的服务如表 7.4 所示。

表 7.4　主要服务

服务	描述
HTTP	封装了 HTTP 请求方法
RequestQptions	封装了 HTTP 请求参数，其具有 Base RequestQptions 子类，默认将请求设置为 GET 方法
ResponseOptions	封装了 HTTP 响应参数，其具有 Base ResponseOptions 子类，默认将响应设置成成功
BrowserXhr	用于创建 XMLHttpRequest 实例
XHRBackend	用于创建一个 XHRConnection 实例，该实例使用 BrowserXhr 对象进行处理

在 Angular 中使用 HttpModule 处理数据，效果如图 7.6 所示。

ID	登录名
4254276	alexpods
1446119	asnowwolf
757236	bradlygreen
3429878	bradrich
782920	calebegg

图 7.6　HttpModule 处理数据

为了实现图 7.6 所示的效果，在 app.module.ts 文件中，导入 HttpModule，代码如 CORE0703 所示。

代码 CORE0703：导入 HttpModule

```
import { BrowserModule } from '@angular/platform-browser';
import { NgModule } from '@angular/core';
import { HttpModule } from '@angular/http'
import { AppComponent } from './app.component';
@NgModule({
```

```
  declarations: [
    AppComponent
  ],
  imports: [
    BrowserModule,
    HttpModule
  ],
  providers: [],
  bootstrap: [AppComponent]
})
export class AppModule { }
```

在组件中导入 HTTP 服务并在组件的构造函数中声明 HTTP 服务。代码如 CORE0704 所示。

代码 CORE0704：处理数据

```
import { Component, OnInit } from '@angular/core';
// 从 @angular/http 模块中导入 Http 类
import { Http } from '@angular/http';
@Component({
})
export class AppComponent implements OnInit {
    // 使用构造函数注入 HTTP 服务
  constructor(private HTTP: Http) { }
  ngOnInit() { }
}
```

在 app.component.ts 文件中设置 HTML 模板，代码如下所示。

```
import { Component, OnInit } from '@angular/core';
// 从 @angular/http 模块中导入 Http 类
import { Http } from '@angular/http';
// 导入 RxJS 中的 map 操作符
import 'rxjs/add/operator/map';
interface Member {
  id: number;
  login: string;
}
  @Component({
```

```
   selector: 'exe-app',
template: `
      <h3>Angular Orgs Members</h3>
      <ul *ngIf="members">
         <li *ngFor="let member of members;">
            <p>
               ID:<span>{{member.id}}</span>
               登录名 : <span>{{member.login}}</span>
            </p>
         </li>
      </ul>
   `
})
export class AppComponent implements OnInit {
   members: Member[];
      // 使用构造函数注入 HTTP 服务
   constructor(private http: Http) { }
   ngOnInit() {  }
}
```

调用 HTTP 服务的 get() 方法，设置请求地址并发送 HTTP 请求。代码如下所示。

```
export class AppComponent implements OnInit {
  members: Member[];
// 使用构造函数注入 HTTP 服务
   constructor(private http: Http) { }
   ngOnInit() {
// 调用 HTTP 服务的 get() 方法，设置请求地址并发送 HTTP 请求（以下地址是网上的）
this.http.get(`https://api.github.com/orgs/angular/members?page=1&per_page=5`)
}
```

调用 Response 对象的 json() 方法，把响应体转成 JSON 对象。代码如下所示。

```
   ngOnInit() {
// 调用 Response 对象的 json() 方法，把响应体转成 JSON 对象
         .map(res => res.json())
         }
```

把请求的结果，赋值给 members 属性。代码如下所示。

```
ngOnInit() {
      .subscribe(data => {
// 把请求的结果，赋值给 members 属性
         if (data) this.members = data;
      });
   }
}
```

快来扫一扫！

提示：经过上面 HTTP 服务的学习，你是否想要了解在 HttpModule 中如何发送 GET 请求，这里有你意想不到的惊喜，心动不如行动，快来扫我吧！

技能点 3 Angular 动画

Angular 的动画系统赋予了制作各种动画效果的能力，致力于构建出与原生 CSS 动画性能相同的动画。以达到用户界面能在不同的状态之间更平滑的转场。其实现首先需要安装项目依赖包，然后引入 BrowserAnimationsModule 函数，最后使用动画触发器（animation triggers），来定义一系列状态和过渡时间。

1 动画触发器的格式

定义一个动画触发器，需要用到 trigger() 方法，这个方法接受两个参数（动画标识符、多个状态转场的数组），定义一个动画触发器格式如下。

```
// triggerName 动画标识符
trigger('triggerName', [
      // 定义的是每个状态的最终样式
      state(' 状态名称一 ', style({CSS 样式 })),
      state(' 状态名称二 ', style({CSS 样式 })),
      // 定义的是状态的转场样式
```

```
    transition(' 状态名称一 => 状态名称二 ',animate(过渡时间)),
    transition(' 状态名称二 => 状态名称一 ', animate(过渡时间))
])
```

根据触发器的格式，其主要由 state、transition 两部分组成，具体介绍如下所示：

（1）state（状态）

state 定义的是每个状态的最终样式，一个触发器里可以有多个 state。例如：飞入动画的两个状态，一是在屏幕外面未飞入状态，另一个是在屏幕中显示的位置。state() 方法接收两个参数：状态名称和 CSS 样式。定义格式如下所示。

```
// triggerName 动画标识符
trigger('triggerName', [
    state( 状态名称 ,style({CSS 样式 }))
])
```

（2）transition（转场）

转场也就是由一个状态过渡到另外一个状态的过程。transition() 方法也接受两个参数：状态之间的切换、animate 属性。定义格式如下所示。

```
// triggerName 动画标识符
trigger('triggerName', [
 transition(' 状态名称一 => 状态名称二 ',animate( 过渡时间 ))
])
```

2　动画触发器的状态

（1）通配符状态

通配符状态可以匹配任意动画状态，当定义那些不需要管当前处于什么状态的样式及转场时使用。如：

- 当该元素的状态从 active（当前状态）变成任意状态（*）时（active → *），动画过渡都会生效。
- 当在任意两个状态之间切换时（* → *），动画过渡都会生效。

（2）void 状态

特殊状态 void 适用于元素未附加到视图（尚未添加或已删除）时的动画。该状态可以定义动画进场和离开。如：

- 进场：void => *。
- 离开：* =>void。

将进场动画与离场动画组合成复合动画。当使用动画时，可以根据当前动画来配置不同状态的进入和离开：

- 不活跃的进场：void=> inactive。

- 主动进场：void=>active。
- 不活跃的离开：inactive=>void。
- 主动离开：active=> void。

注：通配符状态也匹配 void。

3 Angular 动画过渡时间

在 Angular 动画中，转场效果有三种属性：持续时间（duration）、延迟（delay）和缓动（easing）。

（1）持续时间

持续时间控制动画从开始到结束要花多长时间。可以用三种方式定义持续时间：

- 作为一个普通数字，以毫秒为单位，如：100。
- 作为一个字符串，以毫秒为单位，如：'100ms'。
- 作为一个字符串，以秒为单位，如：'0.1s'。

（2）延迟

延迟用于控制动画已触发但尚未真正开始动画之前的时间。将其添加到字符串中的持续时间后面，它的格式和持续时间是一样的，如：

- 等待 100 毫秒，然后运行 200 毫秒：'0.2s 100ms'。

（3）缓动

缓动用于控制动画在运行期间如何加速和减速。比如：使用 ease-in 函数意味着动画开始时相对缓慢，然后在进行中逐步加速。可以通过在字符串中的持续时间和延迟后面添加第三个值来控制使用哪个缓动，如：

- 等待 100 毫秒，然后运行 200 毫秒，并且带缓动效果代码：'0.2s 100ms ease-out'。
- 运行 200 毫秒，并且带缓动效果代码：'0.2s ease-in-out'。

使用 Angular 动画示例代码如下所示。

```
animations: [ // 动画的内容
 trigger('visibilityChanged', [
   // state 控制不同的状态下对应的不同的样式
   state('shown' , style({ opacity: 1, transform: 'scale(1.0)' })),
   state('hidden', style({ opacity: 0, transform: 'scale(0.0)' })),
   // transition 控制状态到状态以什么样的方式来进行转换
   transition('shown => hidden', animate('600ms')),
   transition('hidden => shown', animate('300ms')),
 ])
]
```

4 如何使用 Angular 动画

使用 Angular 动画当页面加载显示黑色，点击状态一切换为黄色，点击状态二切换为蓝色，效果如图 7.7 所示，实现步骤如下所示。

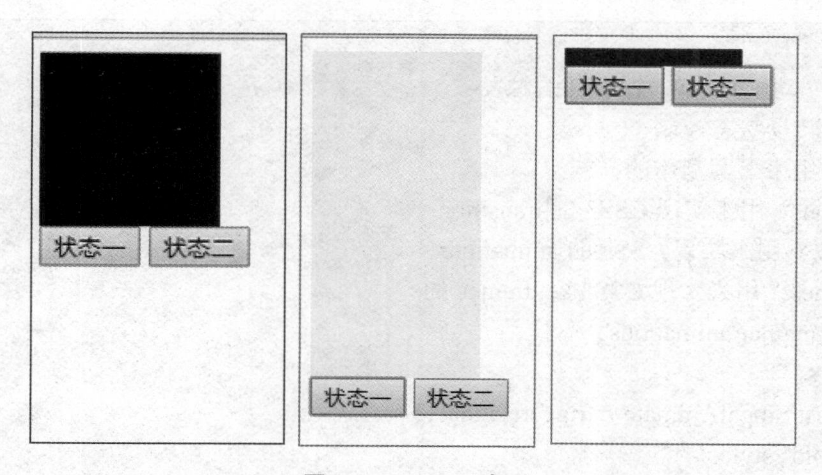

图 7.7　Angular 动画

第一步：安装动画依赖包。

打开命令窗口，输入以下命令，安装项目依赖包，命令如下所示。

```
npm install animations –save
```

第二步：在 AppModule.ts 中引入 BrowserAnimationsModule 函数。

```
import { BrowserAnimationsModule } from '@angular/platform-browser/animations';
```

第三步：在 app.component.html 中设置 HTML 模板，代码如 CORE0705 所示。

代码 CORE0705：HTML 模板

```
<!-- 动画状态占位符 -->
<div class="traffic-light" [@signal]="signal"></div>
<button (click)="go()"> 状态一 </button>
<button (click)="stop()"> 状态二 </button>
```

设置页面加载进来初始化状态，对应的 CSS 代码如 CORE0706 所示。

代码 CORE0706：CSS 代码

```
.traffic-light{
  width: 100px;
  height: 100px;
  background-color: black;
}
```

第四步：新建一个动画的 ts 文件，如 SingalAnimate.ts，定义其动画状态。

```
import {
    trigger, // 动画封装触发，外部的触发器
    state, // 转场状态控制
    style, // 用来书写基本的样式
    transition, // 用来实现 CSS3 的 transition
    animate, // 用来实现 CSS3 的 animations
    keyframes // 用来实现 CSS3 keyframes 的
} from '@angular/animations';

export const SingalAnimate = trigger('signal', [
    state('void', style({
        'transform':'translateX(-100%)'
    })),
    state('go', style({
        'background-color': 'yellow',
        'height':'200px'
})),
    state('stop', style({
        'background-color':'blue',
        'height':'10px'
})),
    transition('void => *', animate(5000)),
    transition('* => *', animate('.5s 1s '))
]);
```

第五步：在组件 ts 里引入动画文件，代码如 CORE0707 所示。

代码 CORE0707：数据服务

```
import { SingalAnimate } from './animate';
@Component({
    selector: 'app-root',
    templateUrl: './app.component.html',
    styleUrls: ['./app.component.css'],
    animations: [
SingalAnimate
]
})
```

第六步：在 app.component.ts 文件编写状态切换方法，代码如 CORE0708 所示。

代码 CORE0708：状态切换方法

```
export class AppComponent {
  signal:string;
  stop(){
    this.signal = 'stop';
  }
  go(){
    this.signal = 'go';
  }
}
```

通过下面八个步骤的操作，实现图 7.2 所示的智慧工厂环安管理模块的效果。

第一步：将环安管理模块分为监控组件、监控详情组件、环安问题列表和会话组件，并创建组件。

第二步：在 ring-management.component.html 文件中对模块进行布局，设置各个组件的渲染位置，代码如 CORE0709 所示。

代码 CORE0709：布局

```
<div style="background-color: white;height: 1100px"><br>
<h2 class="text-center"> 智慧工厂环安管理模块 </h2>
  <div class="row" style="margin-left: 60px">
    <div class="col-md-11">
  <div class="container">
    监控组件
  <div>
    <div> 尾气风机监控模块详情 </div>
    <div> 废气监控详情 </div>
</div>
</div>
    <div> 环安问题列表和会话组件 </div>
</div>
```

第三步：在 app.module.ts 中配置组件路由。引入路由模块，使其点击导航跳转到对应界面，代码如 CORE0710 所示。

代码 CORE0710：路由配置

```
import { NgModule } from '@angular/core';
import { FormsModule } from '@angular/forms';
import { NgbModule } from '@ng-bootstrap/ng-bootstrap';
import { RouterModule, Routes } from '@angular/router';
import { AppComponent } from './app.component';
import {PublicModule} from './public/public.module'
import { RingManagementComponent }
        from './ring-management/ring-management.component'
import {RingManagementModule} from "./ring-management/ring-management.module";
// 路由模块
const appRoutes: Routes = [
  {path:'ringmanagement',component:RingManagementComponent},
];
@NgModule({
  declarations: [
    AppComponent,
    RingManagementComponent,
  ],
  imports: [
    PublicModule,
    BrowserModule,
    FormsModule,
    RouterModule.forRoot(appRoutes),
    NgbModule.forRoot(),
    RingManagementModule,
  ],
  providers: [],
  bootstrap: [AppComponent]
})
export class AppModule { }
```

第四步：设置监控组件。监控组件是对工厂生产中的废水、废气排放等相关数据的采集，本组件主要分为尾气风机监控、合成废水报警、废气监控三个模块，代码如 CORE0711 所示。效果如图 7.8 所示。

代码 CORE0711：监控组件

```
<div class="row pricing-plans plans-4">
  <div class="col-md-3 plan-container best-value">
    <div class="plan">
      <div class="plan-header">
        <div class="plan-title">
          尾气风机监控模块
        </div>
        <div class="plan-price">
          <span class="note"></span><span class="term">2017-09-23</span>
        </div>
        </div>
        <div class="plan-features">
          <ul  class="alert alert-success">
          <li  style="height: 35px;line-height: 10px">
                  <strong> 报警区域：</strong> 工厂三 </li>
          <li style="height: 35px;line-height: 10px">
                  <strong> 监测点名称：</strong> 1 号排口  </li>
          <li style="height: 35px;line-height: 10px">
                  <strong> 工况负荷 ：</strong> 90.0</li>
          <li style="height: 35px;line-height: 10px">
                  <strong> 烟气温度 (℃ )：</strong> 96.0</li>
          <li style="height: 35px;line-height: 10px">
                  <strong> 是否达标：</strong> 是 </li>
          <li style="height: 35px;line-height: 10px">
                  <strong> 含氧量 (%)：</strong> 9.20</li>
        </ul>
      </div>
      <div class="plan-actions" style="text-align: center">
        <label><input type="checkbox"  style="display: none"
  (change)="showHeroes=!showHeroes"> 查看更多 </label>
      </div>
    </div>
  </div>
<!-- 省略部分代码 -->
  </div>
```

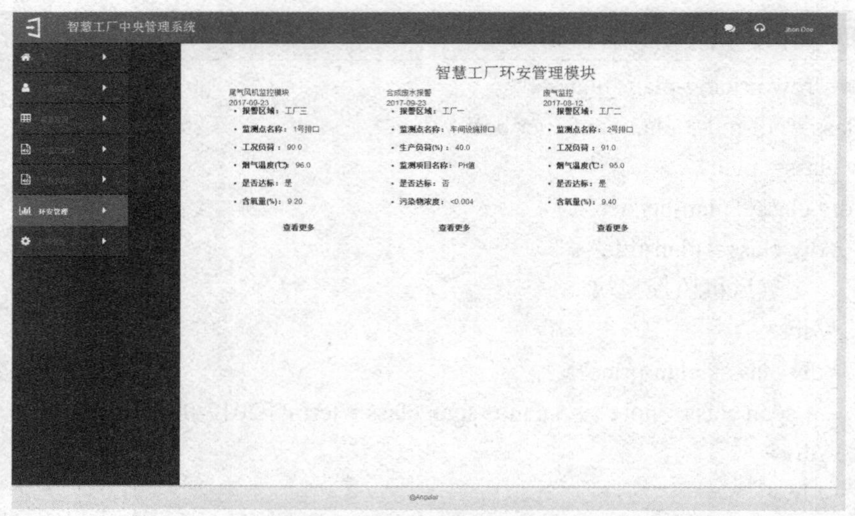

图 7.8 设置样式前

为组件设置布局样式，背景颜色以及更多信息按钮样式。代码如 CORE0712 所示。效果如图 7.9 所示。

代码 CORE0712：CSS 样式

```
.plan-container {
  position: relative;
  float: left;
  margin-bottom: 2em;
}
.plan-container.best-value .plan-header {
  color: #ffffff;
  background-color: #83ae4e;
  text-shadow: 0 -1px 0 rgba(0, 0, 0, 0.25);
}
.plan-container.best-value .plan-price {
  background-color: #94ba65;
}
.plan {
  margin-right: 6px;
  border-radius: 4px;
}
.plan-header {
  text-align: center;
  color: #ffffff;
```

```
    background-color: #676767;
    -webkit-border-top-left-radius: 4px;
    -webkit-border-top-right-radius: 4px;
    -moz-border-radius-topleft: 4px;
    -moz-border-radius-topright: 4px;
    border-top-left-radius: 4px;
    border-top-right-radius: 4px;
    text-shadow: 0 -1px 0 rgba(0, 0, 0, 0.25);
}
.plan-header .plan-title {
    padding: 10px 0;
    font-size: 16px;
    color: #ffffff;
    border-bottom: 1px solid rgba(0, 0, 0, 0.3);
    border-radius: 4px 4px 0 0;
}
.plan-header .plan-price {
    padding: 20px 0 10px;

    font-size: 66px;
    line-height: 0.8em;
    background-color: #797979;
    border-top: 1px solid rgba(255, 255, 255, 0.2);
}
.plan-header .plan-price span.term {
    display: block;
    margin-bottom: 0;
    font-size: 13px;
    line-height: 0;
    padding: 2em 0 1em;
}
.plan-header .plan-price span.note {
    position: relative;
    top: -40px;
    display: inline;
    font-size: 17px;
    line-height: 0.8em;
}
```

图 7.9 设置样式后

第五步：设置监控详情组件。通过给按钮绑定事件，使用 NgIf 指令实现监控组件的显示与隐藏切换。默认隐藏信息，当点击更多信息时，以表格形式弹出监控详情信息。代码如 CORE0713 所示。效果如图 7.10 所示。

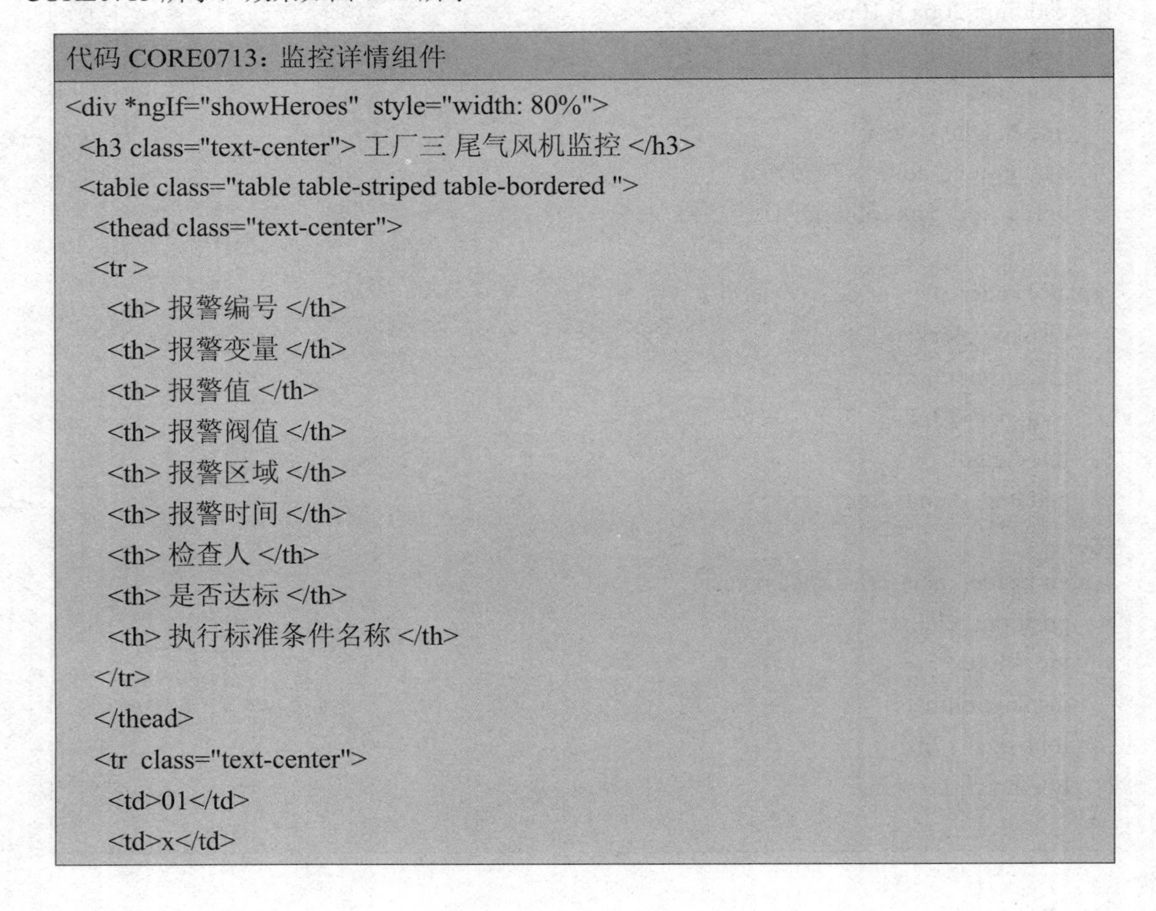

代码 CORE0713：监控详情组件

```
<div *ngIf="showHeroes"  style="width: 80%">
 <h3 class="text-center"> 工厂三 尾气风机监控 </h3>
 <table class="table table-striped table-bordered ">
  <thead class="text-center">
  <tr >
   <th> 报警编号 </th>
   <th> 报警变量 </th>
   <th> 报警值 </th>
   <th> 报警阀值 </th>
   <th> 报警区域 </th>
   <th> 报警时间 </th>
   <th> 检查人 </th>
   <th> 是否达标 </th>
   <th> 执行标准条件名称 </th>
  </tr>
  </thead>
  <tr  class="text-center">
   <td>01</td>
   <td>x</td>
```

```
        <td>54</td>
        <td>64</td>
        <td> 工厂三 </td>
        <td>2017-10-12</td>
        <td> 王敏 </td>
        <td> 是 </td>
        <td> 重点地区锅炉执行的大气污染物特别排放限值 / 燃煤锅炉 </td>
      </tr>
    </tbody>
  </table>
</div>
<!-- 省略部分代码 -->
```

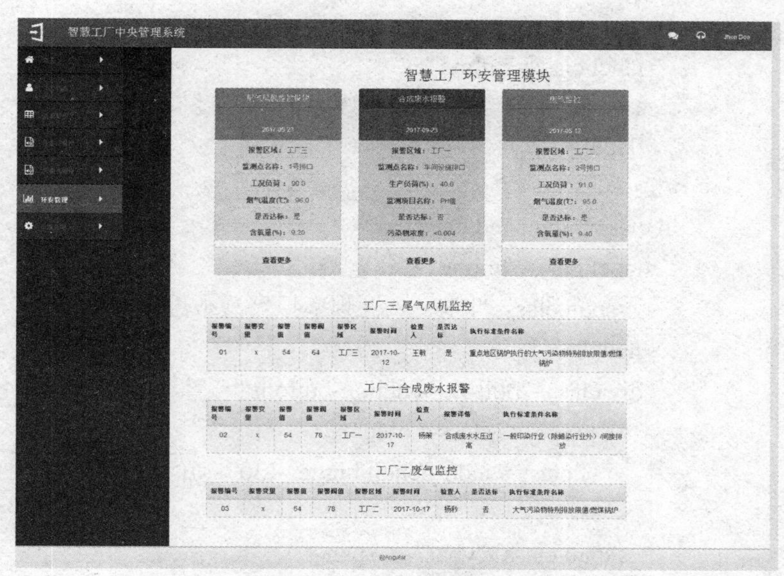

图 7.10　监控详情组件

第六步：设置环安问题列表和会话组件，环安问题列表主要是显示监控出现的一些问题，会话组件为最近联系人的一些对话。通过 Bootstrap 设置其样式，代码如 CORE0714 所示，设置样式前效果如图 7.11 所示。

代码 CORE0714：环安问题列表和会话组件

```
<div class="row" style="width: 80%;margin-bottom: 50px">
  <div class="col-sm-6">
    <div class="widget-box" >
      <div class="widget-header">
        <h4 class="lighter smaller">
```

```
        <i class="icon-rss orange"></i>
        最近
      </h4>
    </div>
    <div class="widget-body">
      <div class="widget-main padding-4">
        <div class="tab-content padding-8 overflow-visible">
          <div id="task-tab" class="tab-pane active">
            <h4 class="smaller lighter green">
              <i class="icon-list"></i>
              环安问题列表
            </h4>
            <ul id="tasks" class="item-list">
              <li class="item-orange clearfix">
                <label class="inline">
                  <span class="lbl"> 合成废水的电流过低 </span>
                </label>
              </li>
              <li class="item-red clearfix">
                <label class="inline">
                  <span class="lbl"> 风机频率报警次数过多 </span>
                </label>
                <div class="pull-right action-buttons">
                  <a href="#" class="blue">
                    <i class="icon-pencil bigger-130"></i>
                  </a>
                  <span class="vbar"></span>
                  <a href="#" class="red">
                  <i class="icon-trash bigger-130"></i>
                  </a>
                  <span class="vbar"></span>
                  <a href="#" class="green">
                    <i class="icon-flag bigger-130"></i>
                  </a>
                </div>
              </li>
              <li class="item-default clearfix">
                <label class="inline">
```

```
            <span class="lbl"> 工业装置排放的有毒气体 </span>
        </label>
        <div class="inline pull-right position-relative dropdown-hover">
            <ul class="dropdown-menu dropdown-only-icon dropdown-yellow
                        dropdown-caret dropdown-close pull-right">
                <li>
                    <a href="#" class="tooltip-success" data-rel="tooltip"
                        title="Mark as done">
                <span class="green">
                    <i class="icon-ok bigger-110"></i>
                </span>
                    </a>
                </li>
                <li>
                    <a href="#" class="tooltip-error" data-rel="tooltip" title="Delete">
            <span class="red">
             <i class="icon-trash bigger-110"></i>
            </span>
                </a>
             </li>
          </ul>
    </div>
 </li>
 <li class="item-blue clearfix">
    <label class="inline">
      <span class="lbl"> 合成循环水不合格 </span>
    </label>
      </li>
      <li class="item-grey clearfix">
         <label class="inline">
            <span class="lbl"> 工业布局不合理 </span>
         </label>
      </li>
      <li class="item-green clearfix">
         <label class="inline">
            <span class="lbl"> 工业装置排放的有毒气体 </span>
         </label>
      </li>
```

```
                        </ul>
                    </div>
                </div>
            </div>
        </div>
    </div>
</div>
</div>
// 省略部分代码
```

图 7.11 设置样式前

设置环安问题列表和会话组件样式，为该组件设置边框、字体样式等，代码如 CORE0715 所示，设置样式后效果如图 7.12 所示。

代码 CORE0715：CSS 代码

```
.widget-box{
padding:0;
-webkit-box-shadow:none;
box-shadow:none;
margin:3px 0;
border-bottom:1px solid #CCC
}
.tab-content.no-padding{
padding:0
```

```
}
.smaller{
line-height:26px
}
.itemdiv{
padding-right:3px;
min-height:66px;
position:relative
}
.user{
display:inline-block;
width:42px;
position:absolute;
left:0
}
.time{
display:block;
font-size:11px;
font-weight:bold;
color:#666;
position:absolute;
ight:9px;top:0
}
```

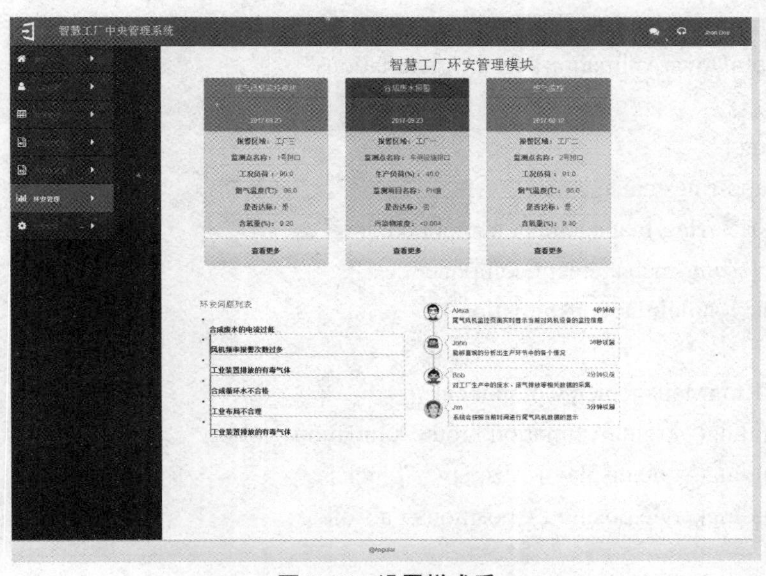

图 7.12　设置样式后

第七步：设置组件加载的动画效果。打开命令窗口输入以下命令，安装动画依赖包，命令如下所示。

```
npm install @angular/animations --save
```

第八步：把 animations 导入到项目中，在 app.module.ts 中导入 BrowserAnimationsModule 模块，并且在 @NgModule 元数据中，把 BrowserAnimationsModule 添加到 imports 列表中，代码如 CORE0716 所示。

代码 CORE0716：导入 animations

```
import { BrowserAnimationsModule } from '@angular/platform-browser/animations';
@NgModule({
  declarations: [
    AppComponent,
  ],
  imports: [
    BrowserAnimationsModule,
  ],
  providers: [],
  bootstrap: [AppComponent]
})
export class AppModule { }
```

导入动画到想要添加的组件的 @Component 元数据中。部分代码 CORE0717 如下。

代码 CORE0717：动画效果

```
import { Component, OnInit,HostBinding } from '@angular/core';
import {slideInDownAnimation} from "./animations";

@Component({
  selector: 'app-ring-management',
  templateUrl: './ring-management.component.html',
  styleUrls: ['./ring-management.component.css'],
  animations: [ slideInDownAnimation ]
})
export class RingManagementComponent  {
  @HostBinding('@routeAnimation') routeAnimation = true;
  @HostBinding('style.display')   display = 'block';
  @HostBinding('style.position')  position = 'absolute';
}
```

至此，智慧工厂环安管理模块制作完成。

本项目通过对智慧工厂环安管理模块的学习，学习了 Angular 服务与动画的基本知识，了解 Angular 动画如何使用，掌握使用 Angular 动画在界面中实现各种动态效果，并具有使用 HTTP 服务与服务器交互的能力，为前端与服务器交互打下基础。

mock　模拟
promise　承诺
injectable　注入
observable　可见的
rxjs　扩展
callback　回调
animation　动画
ease-in-out　加速后然后减少
inactive　不活跃的

一、选择题

1. 下面对服务说法错误的是（　）。
(A)使用单独的服务可以保持组件精简，使其集中精力为视图提供支持
(B)借助模拟(Mock)服务，可以更容易的对组件进行单元测试
(C)由于数据服务总是异步的，因此我们最终会提供一个基于承诺(Promise)的数据服务
(D)Angular 服务不需要任何机制注入，即可完成调用

2. 下面对 Ajax 说法错误的是（　）。
(A)Ajax 是前端项目向服务器请求并操作数据的一种通讯传输技术
(B)处理异步操作的方式有多种，在 Angular 中，常见的是 Observable 处理方式
(C)HTTP 服务的 API 接口返回的也是 Observable 对象
(D)使用 Observable 处理方式可直接发送请求

3. 下面对 JSONP 说法错误的是（　）。
(A)服务端以 callback 函数包裹着 JSON 数据的形式返回一段 JavaScript 代码
(B)实现方式是直接发起 GET 请求并传递 callback 参数给服务器

(C)使用 JSONP，首先在构造函数中注入 JSONP 服务

(D)通过 URLSearchParams 对象构造请求参数，最后调用 JSONP 服务的 get() 方法发起请求

4. 下面对动画说法错误的是(　)。

(A)动画是现代 Web 应用设计中无足轻重

(B)Angular 的动画系统赋予了制作各种动画效果的能力

(C)它们在支持 API 的浏览器中会用原生方式工作

(D)Angular 动画是基于标准的 Web 动画 API(Web Animations API) 构建的

5. 下面对动画函数说法错误的是(　)。

(A)持续时间控制动画从开始到结束要花多长时间

(B)延迟控制的是在动画已经触发但尚未真正开始转场之前要等待多久

(C)可以把延迟控制添加到标签的持续时间后面

(D)缓动函数用于控制动画在运行期间如何加速和减速

二、填空题

1. JSONP 的主要作用是 ____________________________。

2. 对每一个动画转场效果，有三种时间线属性可以调整：________、________ 和 ________。

3. 缓动用于控制动画在运行期间 __________ 和 ____________。

4. Ajax 通过使用 ___________ 可以实现。

5. 持续时间可以用三种方式定义持续时间：__________、__________、__________。

三、上机题

使用 Angular 动画设置符合以下要求。

要求：利用 Angular 动画，实现点击按钮变换颜色，效果如图所示。

项目八　智慧工厂权限管理模块

通过智慧工厂权限管理模块的实现，了解该项目的背景及需求，学习使用 Angular 实现数据的增删改操作功能，掌握其开发所需知识，具有部署测试 Angular 项目的能力。在任务实现过程中：

- 了解智慧工厂项目背景及需求。
- 了解其开发所需知识。
- 学习使用 Angular 实现数据的增删改操作功能。
- 具备部署测试 Angular 项目的能力。

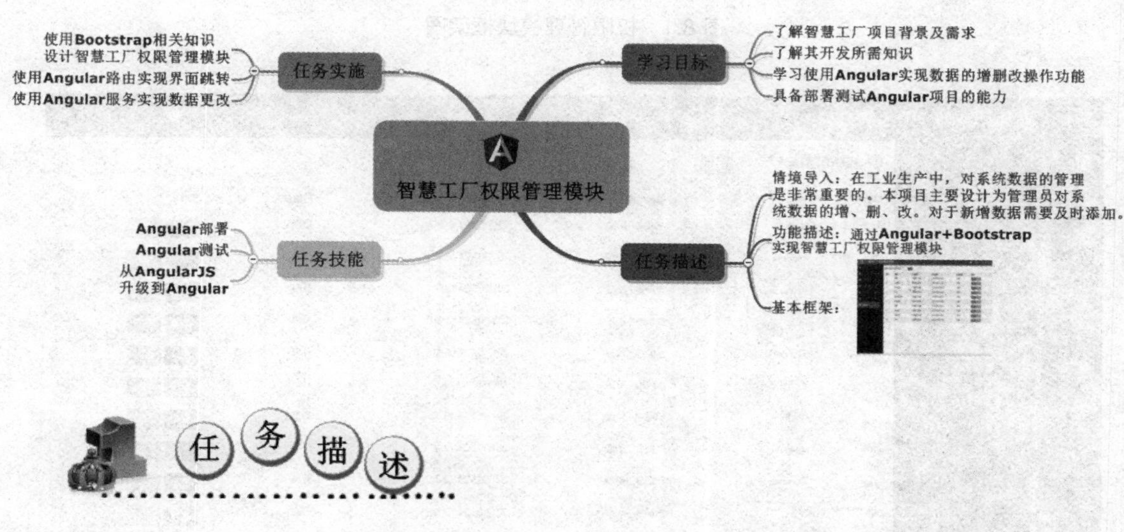

任务描述

【情境导入】

在工业生产中，对系统数据的管理是非常重要的。本项目主要实现管理员对系统数据的增、删、改等操作。当有新数据出现时，需要对数据进行及时更新；当出现错误信息时，可根据编辑功能进行数据修改；而删除功能则主要是为了及时删除过时数据。本项目主要是通过实现智慧工厂的权限管理模块来学习 Angular 的深入知识点。

【功能描述】

使用 Bootstrap+Angular 实现智慧工厂权限管理模块：

- 使用 Bootstrap 相关知识设计智慧工厂权限管理模块。
- 使用 Angular 路由实现界面跳转。
- 使用 Angular 服务实现数据更改。

【基本框架】

基本框架如图 8.1 所示，通过本项目的学习，能将图 8.1 所示的框架图转换成 8.2 的效果图。

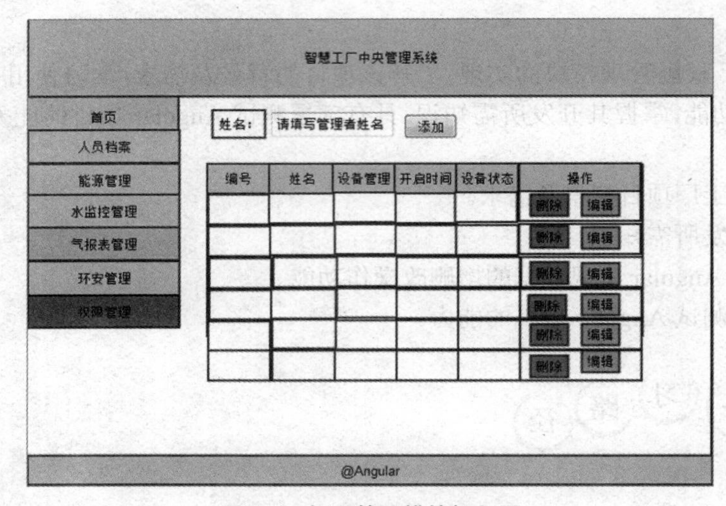

图 8.1 权限管理模块框架图

图 8.2 权限管理模块效果图

技能点 1　Angular 部署

当 Angular 项目完成后，如果想让更多的人来浏览网站，那么就需要将项目部署到服务器上并启动。将 Angular 项目部署到服务器具有多种方式，下面将主要介绍其中的两种方式：简单部署、服务器部署。

1　简单部署

Angular 简单部署指的是将应用直接拷贝到服务器并运行的过程，由于项目在本地已经运行并对出现的 Bug 进行了调试修改，所以可以直接将完整的项目拷贝到能让使用者进行浏览的服务器上，不需要对项目文件进行修改。简单部署特点如下：

- 操作简单：将项目复制到服务器即可。
- 启动步骤多：需要将 Angular 项目启动，之后再将服务器启动。

2　服务器部署

服务器部署是使用服务器启动 Angular 项目，搭建服务器的语言具有很多种，本文用到的是基于 Java 语言（Spring+SpringMVC+MyBatis）搭建的服务器框架，使用 Eclipse 进行后台代码的编写，通过 Tomcat 启动服务器。

Angular 服务器部署的步骤如下所示。

第一步：打开控制台进入项目路径，运行如下代码打包项目，打包后的效果如图 8.3 所示。

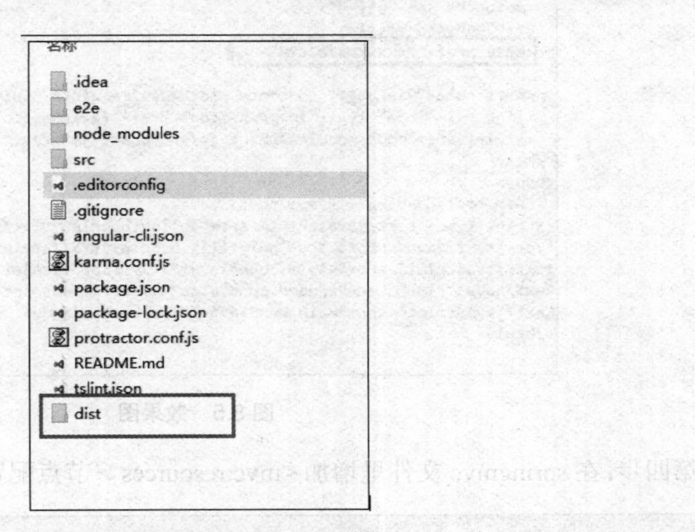

图 8.3　效果图

```
npm run build 或 ng build
```

第二步：拷贝 dist 文件夹到 Eclipse 项目文件里，效果如图 8.4 所示。

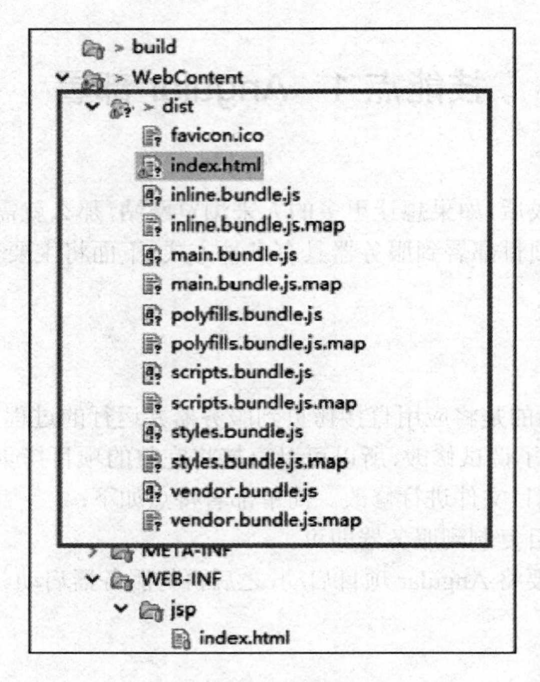

图 8.4 效果图

第三步：修改 dist/index.html 文件下的 <base href=""> 为项目名 + 文件名，效果如图 8.5 所示。

```
<!doctype html>
<html>
<head>
  <meta charset="utf-8">
  <title>MyApp</title>
  <base href="/xxxxxxx/dist/">

  <meta name="viewport" content="width=device-width, initial-scale=1">
  <link rel="icon" type="image/x-icon" href="favicon.ico">
  <script src="node_modules/chart.js/src/chart.js"></script>
</head>
<body>
  <app-root>Loading...</app-root>
<script type="text/javascript" src="inline.bundle.js"></script><script
type="text/javascript" src="polyfills.bundle.js"></script><script type="
text/javascript" src="styles.bundle.js"></script><script type="
text/javascript" src="vendor.bundle.js"></script><script type="
text/javascript" src="main.bundle.js"></script></body>
</html>
```

图 8.5 效果图

第四步：在 springmvc 文件里增加 <mvc:resources > 节点配置。

```
<mvc:resources location="/dist/" mapping="/dist/**" />
```

第五步：在浏览器运行 http://localhost:8080/xxxxxxx/dist/index.html。

快来扫一扫！

提示：当你学到这里，是否有收获呢？在以后的工作中，少说空话多做工作，扎扎实实埋头苦干，才会有不一样的收获。扫描图中二维码，也将有更多的收获。

技能点 2　Angular 测试

1　Angular 测试简介

当设计项目时，难免会在项目设计、程序出现错误，通过测试可以及时为程序发出警告，且测试程序也可起到澄清代码的作用，无论是代码被正确使用还是错误使用。在一定程度上，测试能够减少程序后期错误率。目前，可以通过多种工具和技术来编写、运行 Angular 测试程序。部分工具、技术如表 8.1 所示。

表 8.1　Angular 测试程序技术 / 工具

技术 / 工具	描　述
Jasmine	Jasmine 测试框架提供了所有编写基本测试的工具。自带 HTML 测试运行器，用来在浏览器中执行测试程序
Angular 测试工具	Angular 测试工具为被测试的 Angular 应用代码创建测试环境。在应用代码与 Angular 环境互动时，使用 Angular 测试工具来限制和控制应用的部分代码
Karma	karma 测试运行器是在开发应用的过程中，编写和运行单元测试的理想工具
Protractor	使用 Protractor 来编写和运行端对端（e2e）测试程序。在端对端测试中，一条进程运行真正的应用，另一条进程运行 Protractor 测试程序，模拟用户行为，判断应用在浏览器中的反应是否正确

2　Karma 测试

Karma 是 Testacular 的新名字，在 2012 年 Google 开源了 Testacular，2013 年 Testacular 改名为 Karma。Karma 是一个基于 Node.js 的管理工具。该工具可用于测试所有主流 Web 浏览

器，也可集成到 CI（Continuous integration）工具，或与其他代码编辑器一起使用。其具有可以监控文件变化，通过自行执行，以 console.log 显示测试结果的强大特性。

Karma 主要测试的是后缀名为 .spec.ts 的文件。首先需要创建测试文件，在项目的根目录下创建 xx.spec.ts（后缀必须是 .spec.ts）文件，接下来编写 Jasmine 测试程序，示例代码如下所示。

```
describe('xx tests', () => {
  it('true is true', () => expect(true).toBe(true));
});
```

将测试代码放在 app 文件夹下 xx.spec.ts 中，karma.conf.js 文件将指引 Karma 在这个文件夹中寻找测试程序 .spec.ts 文件。测试过程如下。

（1）运行 Karma

在命令行编写以下命令，在 Karma 中运行测试程序。效果如图 8.6 所示。

```
npm test
```

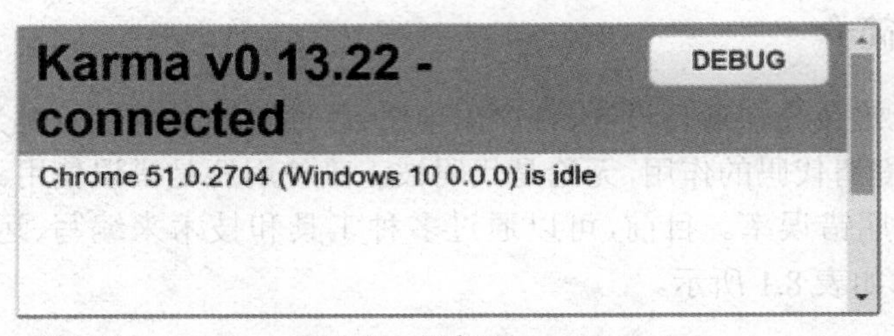

图 8.6　运行 Karma

控制台输出代码如下所示。

```
> npm test
...
[0] 1:37:03 PM - Compilation complete. Watching for file changes.
...
[1] Chrome 51.0.2704: Executed 0 of 0 SUCCESS
     Chrome 51.0.2704: Executed 1 of 1 SUCCESS
SUCCESS (0.005 secs / 0.005 secs)
```

编译器和 Karma 都会持续运行。编译器的输入信息前面有 [0]，Karma 的输出前面有 [1]。将期望从 true 变换为 false。编译器监视器检测到这个变化并重新编译。

```
[0] 1:49:21 PM - File change detected. Starting incremental compilation...
[0] 1:49:25 PM - Compilation complete. Watching for file changes.
```

Karma 监视器检测到编译器输出的变化，并重新运行测试。

```
[1] Chrome 51.0.2704 1st tests true is true FAILED
[1] Expected false to equal true.
[1] Chrome 51.0.2704: Executed 1 of 1 (1 FAILED) (0.005 secs / 0.005 secs)
```

当从 false 恢复为 true。两个进程都检测到这个变化，自动重新运行，Karma 报告测试成功。

（2）调试测试程序

在浏览器中，点击键盘 F12 按钮或鼠标右击点击检查按钮，打开 Sources 选项，选择测试文件，像调试应用一样调试测试程序 spec。具体步骤如下所示。效果如图 8.7 所示。

步骤 1　显示 Karma 的浏览器窗口。

步骤 2　点击“重新加载”按钮，它打开一页新浏览器标签并重新开始运行测试程序。

步骤 3　在浏览器页面点击键盘 F12 按钮或鼠标右击点击检查按钮。

步骤 4　选择“Sources”。

步骤 5　打开 test.spec.ts 测试文件。

步骤 6　在测试程序中设置断点。

步骤 7　刷新浏览器 ... 然后它就会停在断点上。

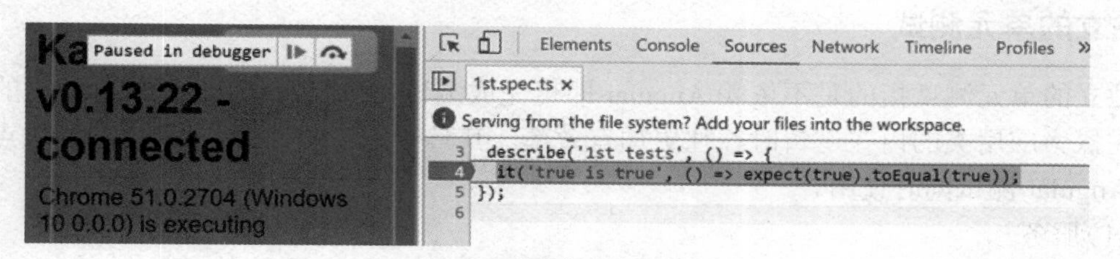

图 8.7　调试测试程序

3　测试组件

组件是构成 Angular 项目的基础，因此，大部分编程人员将会首先测试组件，示例代码如下所示。

```
import { Component } from '@angular/core';
@Component({
  selector: 'app-root',
  template: '<h1>{{title}}</h1>',
})
export class AppComponent {
  title = 'Angular';
}
```

在测试文件中，需要 import 语句来引入测试所需符号。

```
describe('AppComponent (inline template)', () => {
  let comp:    AppComponent;
  let fixture: ComponentFixture< AppComponent >;
  let de:      DebugElement;
  let el:      HTMLElement;
  beforeEach(() => {
    TestBed.configureTestingModule({
      declarations: [AppComponent],
    });
    fixture = TestBed.createComponent(BannerComponent);
    comp = fixture.componentInstance; // 声明测试组件
    // 查询标题
    de = fixture.debugElement.query(By.css('h1'));
    el = de.nativeElement;
  });
});.
```

4 独立的单元测试

独立的单元测试指的是不依赖 Angular 即可完成测试，相比较 Angular 测试，独立的单元测试在探索应用类的内在逻辑时往往更加有效率。其具有更易阅读、编写和维护等优点。且能和 Angular 测试同时使用。

（1）服务

通过服务可应用单元测试，使用单元测试可以缩减配置和代码的复杂性，示例代码如下所示。

```
// 没有 import 导入
describe('FancyService without the TestBed', () => {
  let service: FancyService;
  beforeEach(() => { service = new FancyService(); });
  it('#getValue should return real value', () => {
    expect(service.getValue()).toBe('real value');
 });
 it('#getAsyncValue should return async value', (done: DoneFn) => {
  service.getAsyncValue().then(value => {
    expect(value).toBe('async value');
    done();
  });
 });
```

```
});
  it('#getAsyncValue should return async value', (done: DoneFn) => {
    service.getAsyncValue().then(value => {
      expect(value).toBe('async value');
      done();
    });
  });
  it('#getTimeoutValue should return timeout value',  (done: DoneFn) => {
      service = new FancyService();
      service.getTimeoutValue().then(value => {
          expect(value).toBe('timeout value');
          done();
      });
  });
  it('#getObservableValue should return observable value', (done: DoneFn) => {
      service.getObservableValue().subscribe(value => {
          expect(value).toBe('observable value');
          done();
      });
  });
});
```

（2）带依赖的服务

服务有时会依赖其他服务，Angular 通过构造函数依赖注入它们。在许多情况下，创建和手动注入依赖来的更加容易。示例代码如下所示。

```
@Injectable()
export class DependentService {
  constructor(private dependentService: FancyService) { }
  getValue() { return this.dependentService.getValue(); }
}
```

通过 getValue() 方法对依赖注入的 FancyService 服务进行操作。以下为几种测试它的方法。

```
describe('DependentService without the TestBed', () => {
  let service: DependentService;
// 使用 new 创建 FancyService 实例，并将它传递给 DependentService 构造函数
  it('#getValue should return real value by way of the real FancyService', () => {
```

```
    service = new DependentService(new FancyService());
    expect(service.getValue()).toBe('real value');
});

it('#getValue should return faked value by way of a fakeService', () => {
    service = new DependentService(new FakeFancyService());
    expect(service.getValue()).toBe('faked value');
});

it('#getValue should return faked value from a fake object', () => {
    const fake =  { getValue: () => 'fake value' };
    service = new DependentService(fake as FancyService);
    expect(service.getValue()).toBe('fake value');
});

it('#getValue should return stubbed value from a FancyService spy', () => {
    const fancy = new FancyService();
    const stubValue = 'stub value';
    const spy = spyOn(fancy, 'getValue').and.returnValue(stubValue);
    service = new DependentService(fancy);

    expect(service.getValue()).toBe(stubValue, 'service returned stub value');
    expect(spy.calls.count()).toBe(1, 'stubbed method was called once');
    expect(spy.calls.mostRecent().returnValue).toBe(stubValue);
  });
});
```

（3）管道

管道在 Angular 项目中非常容易测试，无需 Angular 测试工具。管道类通过 transform() 方法转换输入值到输出值。transform() 方法的实现几乎不与 DOM 交互。除了 @Pipe 元数据和接口外，大部分管道不依赖 Angular。示例代码如下所示。

```
import { Pipe, PipeTransform } from '@angular/core';

@Pipe({name: 'titlecase', pure: false})
// 转换为标题大小写：将字符串中单词的第一个字母大写
export class TitleCasePipe implements PipeTransform {
  transform(input: string): string {
```

```
        return input.length === 0 ? '' :
          input.replace(/\w\S*/g, (txt => txt[0].toUpperCase() + txt.substr(1).toLowerCase() ));
    }
}
```

以下为使用单元测试对正则表达式的类进行测试。

```
describe('TitleCasePipe', () => {
  let pipe = new TitleCasePipe();

  it('transforms "abc" to "Abc" ', () => {
    expect(pipe.transform('abc')).toBe('Abc');
  });

  it('transforms "abc def" to "Abc Def" ', () => {
    expect(pipe.transform('abc def')).toBe('Abc Def');
  });
});
```

（4）与 Angular 测试同时使用

有些管道的测试程序是孤立的，不能验证 TitleCasePipe（前文的管道中用到）应用到组件上时是否工作正常，因此添加 Angular 测试。示例代码如下所示。

```
it('should convert hero name to Title Case', () => {
  const inputName = 'quick BROWN  fox';
  const titleCaseName = 'Quick Brown  Fox';

  // 模拟用户输入新名称进入输入框
  page.nameInput.value = inputName;

  // 分派一个 DOM 事件
  page.nameInput.dispatchEvent(newEvent('input'));

  fixture.detectChanges();
  expect(page.nameDisplay.textContent).toBe(titleCaseName);
});
```

（5）组件

组件测试通常检查该组件类是如何与模板或者其他合作组件交互。

```
@Component({
  selector: 'button-comp',
  template: '
    <button (click)="clicked()">Click me!</button>
    <span>{{message}}</span>'
})
export class ButtonComponent {
  isOn = false;
  clicked() { this.isOn = !this.isOn; }
  get message() { return 'The light is ${this.isOn ? 'On' : 'Off'}'; }
}
```

下面的 Angular 测试演示点击模板里的按钮后，进行实时更新。

```
it('should support clicking a button', () => {
  const fixture = TestBed.createComponent(ButtonComponent);
  const btn  = fixture.debugElement.query(By.css('button'));
  const span = fixture.debugElement.query(By.css('span')).nativeElement;

  fixture.detectChanges();
  expect(span.textContent).toMatch(/is off/i, 'before click');

  click(btn);
  fixture.detectChanges();
  expect(span.textContent).toMatch(/is on/i, 'after click');
});
```

该判断验证了数据绑定从一个 HTML 控件（<button>）流动到组件，以及从组件回到不同的 HTML 控件（<span>）。通过测试程序说明组件和它的模块是否设置正确。单元测试可以更快的在 API 边界探测组件。示例代码如下所示，功能是用来验证面对多种输入时组件的输出。

```
describe('ButtonComp', () => {
  let comp: ButtonComponent;
  beforeEach(() => comp = new ButtonComponent());

 it('#isOn should be false initially', () => {
  expect(comp.isOn).toBe(false);
 });
```

```
it('#clicked() should set #isOn to true', () => {
  comp.clicked();
  expect(comp.isOn).toBe(true);
});
it('#clicked() should set #message to "is on" ', () => {
  comp.clicked();
  expect(comp.message).toMatch(/is on/i);
});

it('#clicked() should toggle #isOn', () => {
  comp.clicked();
  expect(comp.isOn).toBe(true);
  comp.clicked();
  expect(comp.isOn).toBe(false);
});
});
```

技能点 3　从 AngularJS 升级到 Angular

AngularJS 指的是 Angular 的所有 v1.x 版本，而 Angular 则是 v2.x 及其以后的所有版本。AngularJS 与 Angular 绝大部分不相同，AngularJS 采用的是 JavaScript 编写，而 Angular 采用的是 TypeScript 编写。Angular 与 AngularJS 相比，整体性能得到了很大的提升，成本也降低许多。因此，部分编程人员想将 AngularJS 项目升级到 Angular 项目。但是对于大型 AngularJS 应用，在决定升级到 Angular 之前，需要对业务进行深入思考。以下为升级需要做的工作。

1　升级准备工作

（1）AngularJS 规则

Angular 保留 AngularJS 中性能好的部分。有一些特别的规则可以让使用 Angular 的 upgrade/static 模块进行增量升级变得更简单：

- 单一规则：规定每个文件应该只放一个组件。
- 按特性分目录的结构和模块化规则：应用程序中的不同部分应该被分到不同的目录和 Angular 模块中。

（2）使用模块加载器

使用模块加载器，如：SystemJS、Webpack 或 Browserify，通过模块加载器可以在程序中使用 TypeScript 等。

（3）使用 TypeScript

在 AngularJS 中引入 TypeScript 编译器。

- 对使用模块加载器的程序，TypeScript 的导入和导出语法可以把代码组织成模块。
- 可以逐步把类型注解添加到现有函数和变量上，以固定它们的类型。
- TypeScript 中新增的特性，如箭头函数、解构赋值等可添加进 AngularJS 中。
- 服务和控制器可以转成类。

2 升级过程

在升级过程中，ngUpgrade 库是非常重要的，通过 ngUpgrade 库可在同一个项目中混用并匹配 AngularJS 和 Angular 的组件，做到无缝的互操作（不同平台或编程语言之间交换和共享数据的能力）。通过 upgrade 模块导入 UpgradeModule 服务，可实现一个框架中管理的组件和服务能和来自另一个框架的进行互操作。通过依赖注入、DOM 和变更检测等可体现。如：依赖注入。

依赖注入是 AngularJS 与 Angular 共同存在的性能，但在工作原理上具有一定的区分。如表 8.2 所示。

表 8.2　AngularJS 与 Angular 工作原理区分

AngularJS	Angular
Token 永远是字符串	Token 可能有不同的类型
只有一个注入器	有一个根注入器，而且每个组件也有一个自己的注入器

通过 UpgradeModule 服务，可解决工作原理上的不同，使其无缝的对接。通过升级 AngularJS，可将 AngularJS 中被注入的服务在 Angular 的代码中可用。

（1）开始用 TypeScript

Angular 使用 TypeScript 编写，TypeScript 是 JavaScript 的超集，带有更多的特性。在编写项目时，代码最终都会编译成 ts 文件。示例代码如下所示。

```
// 对象属性增强
var exports = {
  search: search,
  setType: setType,
  setDuration: setDuration
};
// 可以写成这样
var exports = {
  search,
  setType,
  setDuration
```

```
};
// 使用"胖箭头"=> 可以简化代码并增强可读性
var videoIds = response.data.items.map(function(video){
  return video.id[idPropertyName[activeType]];
}).join(',');
// 使用了"胖箭头"符号
var videoIds = response.data.items.map((video) => {
  return video.id[idPropertyName[activeType]];
}).join(',');
```

（2）使用".service"文件替换".factory"文件

在 Angular 中，通过 TypeScript 可以使用 class（关键字）来创建新对象甚至扩展其他对象，而在 AngularJS 中使用普通函数即可完成调用，不需要 class（关键字）。因此需要将 AngularJS 代码中的 factories 转化成 services，并且使用 class（关键字）替代 function，代码如下所示。

```
(function() {
  'use strict';
  /* @ngInject */
  class YoutubePlayerApi {
    /* @ngInject */
    constructor ($window, $q) {
      /*jshint validthis: true */
      this.deferred = $q.defer();
      // 当 API 准备好时，Youtube 回调
      $window.onYouTubeIframeAPIReady = () => {
        this.deferred.resolve()
      };
    }
    // 注入 YouTube 的 iFrame API
    load () {
      let validProtocols = ["http:","https:"];
      let url = "//www.youtube.com/iframe_api";
      // 我们愿意使用相关的 url 协议，但为避免协议不可用，还是回退到 'http:'
      if (validProtocols.indexOf(window.location.protocol) < 0) {
        url = "http:" + url;
      }
      let tag = document.createElement'script';
      tag.src = url;
```

```
            let firstScriptTag = document.getElementsByTagName("script")[0];
            firstScriptTag.parentNode.insertBefore(tag, firstScriptTag);
            return this.deferred.promise;
        }
    }
    angular
        .module("youtube.player")
        .service("YoutubePlayerApi", YoutubePlayerApi);
})();
```

（3）编写 Controllers 时使用“类”替换“函数”

在 Angular 中几乎不适用控制器，且 Angular 基于组件，每个组件通过类来控制，而在 AngularJS 中控制器总是不停的变化，每个指令都对应一个 controller 函数来控制。当进行升级时，把 controller 写成类，会使代码升级到 Angular 组件变得更容易，示例代码如下所示。

```
class DurationCtrl {
    constructor (YoutubeSearch) {
        this.YoutubeSearch = YoutubeSearch;
        this.durations = [
            'Any',
            'Short (less then 4 minutes)',
            'Medium (4-20 minutes)',
            'Long (longer than 20 minutes)'
        ];
        this.durationsMap = [
            ' ',
            'short',
            'medium',
            'long'
        ];
    }
    onDurationChange (duration, index) {
        this.YoutubeSearch.setType(this.YoutubeSearch.types.VIDEO);
        this.YoutubeSearch.setDuration(this.durationsMap[index]);
        this.YoutubeSearch.search();
    }
}
angular
```

```
  .module('echoes')
  .controller('DurationCtrl', DurationCtrl);
```

（4）使用指令封装代码

不能在 index.html 或任何未关联指令的模板中编写任何 Angular 代码。如：在 AngularJS 列表中使用 ng-repeat 循环遍历数组时，可通过创建指令“<profile-cards>”对代码进行封装。示例代码如下所示。

```
<div ng-repeat="person in vm.persons">
  <img ng-src="person.thumb">
  <h3>{{:: person.name }}</h3>
  <aside>{{:: person.moto }}</aside>
  <p>
    {{:: person.description }}
  </p>
</div>
<!-- 可以转换成一个组件 -->
<div ng-repeat="person in vm.persons">
  <person-profile-card model="person"></person-profile-card>
</div>
<!-- 可以成为另一个列表组件 -->
<profile-cards items="vm.persons"></profile-cards>
```

（5）使用 ng-upgrade

Angular 采用了简单、大方的语法来定义组件、指令。为了体验 Angular 的组件语法，可以在 AngularJS 应用中使用 Angular 的 TypeScript 标准语法来定义一个指令，示例代码如下所示。

```
var myApp = ng
  .Component({
    selector: 'youtube-videos'
    providers: [
      'core.services'
    ],
    bindings: {
      videos: '@'
    }
  })
  .View({
```

```
    templateUrl: 'app/youtube-videos/youtube-videos.tpl.html'
  })
  .Class({
    constructor: 'YoutubeVideosCtrl'
  })
```

以下使用相当于 Angular JS 指令，代码如下所示。

```
angular.module('youtube-videos', [
        'core.services'
    ])
    .directive('youtubeVideos', youtubeVideos);

  function youtubeVideos () {
    var directive = {
      controller: 'YoutubeVideosCtrl',
      controllerAs: 'vm',
      restrict: 'E',
      replace: true,
      template: 'app/youtube-videos/youtube-videos.tpl.html',
      bindToController: true,
      scope: {
        videos: '@'
      }
    };
    return directive;
  }
```

快来扫一扫！

提示：在前端开发中，会有很多的框架。当我们学会了 Angular 前端框架后，是否好奇 Angular 与其他前端框架的区别？扫描图中二维码，会有你想不到的惊喜。

通过下面十个步骤的操作，实现图 8.2 所示的智慧工厂权限管理模块的效果。

第一步：新建权限管理组件以及查看权限详情组件。权限管理模块为一个表格，可以对表格数据进行增删改操作。

第二步：配置路由。点击跳转到权限详情组件时使其携带 id。代码如 CORE0801 所示。

代码 CORE0801：app.modules.ts

```
import { RouterModule, Routes } from '@angular/router';

const appRoutes: Routes = [
  { path: 'heroesgas', component: HeroesgasComponent },
  { path: 'heroDetailwg/:id', component: HeroDetailwgComponent },
];

@NgModule({
  imports: [
    RouterModule.forRoot(appRoutes),
  ],
})
```

第三步：在 heroesgas.component.html 文件中，编写权限管理组件模板，使其以表格形式显示展示数据。定义数据变量，表格数据使用 NgFor 指令双向绑定显示。代码如 CORE0802 所示。

代码 CORE0802：在 heroesgas.component.html 文件中对模块进行布局

```
<div style="background-color: white;height: 900px">
  <h1></h1>
  <div>
    <label> 姓名 :</label> <input #heroName placeholder=" 请填写管理者姓名 "
              style="border-radius: 5px"/>
    <button class="btn btn-danger">
      添加
    </button>
  </div>
  <table class="table table-bordered table-striped" style="margin-top: 20px">
    <tr *ngFor="let hero of heroes ">
      <th> 编号 </th>
```

```
        <th> 姓名 </th>
        <th> 设备管理 </th>
        <th> 开启时间 </th>
        <th> 设备状态 </th>
        <th> 操作 </th>
      </tr>
      <tr>
        <td>{{hero.id}}</td>
        <td>{{hero.name}}</td>
        <td>{{hero.Equipmentmanagement}}</td>
        <td>{{hero.Openingtime}}</td>
        <td>{{hero.devicestatus}}</td>
        <td><button class="btn btn-primary"> 删除 </button>
          <button class="btn btn-success"> 编辑 </button>
        </td>
      </tr>
  </table>
</div>
```

在 app 目录下新建 hero.ts 文件，用于声明人员属性。代码如 CORE0803 所示。

代码 COR0803：hero.ts

```
export class Hero {
id: number;
name: string;
Equipmentmanagement:string; // 设备管理
Openingtime:string;// 开启时间
devicestatus:string// 设备状态
}
```

第四步：在 app.modules.ts 文件中导入 HttpModule 模块。HttpModule 保存着 HTTP 相关服务提供商的全集。

```
import { NgModule }      from '@angular/core';
import { BrowserModule } from '@angular/platform-browser';
import { FormsModule }   from '@angular/forms';
import { HttpModule }    from '@angular/http';

import { AppRoutingModule } from './app-routing.module';
```

```
import { AppComponent }         from './app.component';
import { DashboardComponent }   from './dashboard.component';
import { FooterComponent } from './footer/footer.component';
import { NavbarComponent } from './navbar/navbar.component';

@NgModule({
  imports: [
    BrowserModule,
    FormsModule,
    HttpModule,
    AppRoutingModule
  ],
  declarations: [
    AppComponent,
    DashboardComponent,
    FooterComponent,
    NavbarComponent,
  ],
  providers: [],
  bootstrap: [ AppComponent ]
})
export class AppModule { }
```

第五步：在 app 目录下新建 in-memory-data.service.ts（文件名可变）文件用来存放表格数据。在 createDb() 方法中新建变量存放数组信息。代码如 CORE0804 所示。

代码 CORE0804：in-memory-data.service.ts 文件用来存放数据

```
import { InMemoryDbService } from 'angular-in-memory-web-api';
export class InMemoryDataService implements InMemoryDbService {
  createDb() {
    const heroes2 = [
      {
  id: 1,
  name: ' 肖峰 ',
  Equipmentmanagement: ' 气报表一阶段 ',
  Openingtime: '2017/05/26 13:30',
  devicestatus:' 维修 '
```

```
},
{
  id: 2,
  name: 'Magma',
  Equipmentmanagement: ' 气报表二阶段 ',
  Openingtime: '2017/03/24 12:30',
  devicestatus:' 开启 '
},
      // 省略部分代码
    ];
      return { heroes2};
  }
}
```

第六步：创建 Hero2Service.ts 服务，导入 Hero、Http 以及 RxJS 库等声明文件，并写入 getHeroes() 方法，通过 HTTP 来获取数据。代码如 CORE0805 所示。

代码 CORE0805：Hero2Service.ts 服务文件

```
import { Injectable }    from '@angular/core';
import { Headers, Http } from '@angular/http';
// 从 RxJS 库中导入操作符，Angular 中的 observable 对象并没有 topromise 操作符，需
// 要借助其他工具
import 'rxjs/add/operator/toPromise';

import { Hero } from './hero';
@Injectable()
export class Hero2Service {
  private heroesUrl = 'api/heroes';
  private headers = new Headers({'Content-Type': 'application/json'});
  constructor(private http: Http) { }

  getHeroes(): Promise<Hero[]> {
// Angular 的 http.get 返回一个 RxJS 的 Observable 对象。
//Observable（可观察对象）是一个管理异步数据流的强力方式
    return this.http.get(this.heroesUrl)
// 利用 toPromise 操作符把 Observable 直接转换成 Promise 对象
               .toPromise()
// 在 promise 的 then() 回调中，调用 HTTP 的 Reponse 对象的 json 方法，以提取出其中
// 的数据
```

```
                    .then(response => response.json().data as Hero[])
// catch 了服务器的失败信息，并把它们传给了错误处理器：
                    .catch(this.handleError);
  }
private handleError(error: any): Promise<any> {
      console.error('An error occurred', error);
      return Promise.reject(error.message || error);
  }
}
```

第七步：在 Hero2Service.ts 服务文件中，添加 getHero() 方法来发起一个通过 id 获取数据的请求。此方法基本与 getHeroes() 方法一致，通过在 URL 中添加人员的 id 来告诉服务器应该获取哪个人员的信息。代码如 CORE0806 所示。

代码 CORE0806：Hero2Service.ts 服务文件

```
getHero(id: number): Promise<Hero> {
  const url = `${this.heroesUrl}/${id}`;
  return this.http.get(url)
     .toPromise()
// 把响应中返回的 data 改为一个英雄对象，而不是对象数组
     .then(response => response.json().data as Hero)
     .catch(this.handleError);
}
```

第八步：实现添加功能。在 heroesgas.component.html 文件中，管理员对信息有添加功能。给添加按钮绑定 click 事件。代码如下所示。

```
<button (click)="add(heroName.value); " class="btn btn-danger"> 添加 </button>
```

在 heroes.component.ts 中，添加 add() 方法。当点击事件触发时，调用组件的点击处理器，然后清空这个输入框，以便用来输入另一个名字。代码如 CORE0807 所示。

代码 CORE0807：heroes.component.ts

```
add(name: string): void {
  name = name.trim();
  if (!name) { return; }
  this.heroService.create(name)
     .then(hero => {
       this.heroes.push(hero);
       this.selectedHero = null;
```

```
    });
}
```

当指定的名字不为空的时候，点击处理器委托 hero 服务来创建一个具有此名字的人员，并把新的人员添加到数组中。在 Hero2Service 类中实现 create() 方法。代码如 CORE0808 所示。

代码 CORE0808：Hero2Service.ts

```
create(name: string): Promise<Hero> {
  return this.http
    .post(this.heroesUrl, JSON.stringify({name: name}), {headers: this.headers})
    .toPromise()
    .then(res => res.json().data as Hero)
    .catch(this.handleError);
}
```

第九步：实现删除功能。在 heroesgas.component.html 文件中，管理员对信息有删除功能。给删除按钮绑定 click 事件。代码如下所示。

```
<button (click)="delete(hero); " class="btn btn-primary"> 删除 </button>
```

在 heroes.component.ts 中添加 delete() 方法。代码如 CORE0809 所示。

代码 CORE0809：heroes.component.ts

```
delete(hero: Hero): void {
  this.heroService
    .delete(hero.id)
    // 点击处理器还应该阻止点击事件向上冒泡
    .then(() => {
      this.heroes = this.heroes.filter(h => h !== hero);
      if (this.selectedHero === hero) { this.selectedHero = null; }
    });
}
```

在 hero2.service.ts 中把删除人员的操作委托给了 HTTP 服务的 delete() 方法，使用 HTTP 的 delete() 方法来从服务器上移除该信息，并在网页中更新显示。代码如 CORE0810 所示。效果如图 8.8 所示。

代码 CORE0810：heros2.service.ts

```
delete(id: number): Promise<void> {
  const url = `${this.heroesUrl}/${id}`
```

```
    return this.http.delete(url, {headers: this.headers})
      .toPromise()
      .then(() => null)
      .catch(this.handleError);
}
```

图 8.8　权限管理组件

第十步：在 hero-detail.component.html 中。当点击编辑按钮跳转到人员信息详细信息页面，可以对信息进行修改。在权限详情模板（hero-detail.component.html）的底部添加一个返回和保存按钮，并绑定了一个 click 事件。代码如 CORE0811 所示。

代码 CORE0811：hero-detail.component.html

```
<div style="background-color: white;height: 900px">
  <h2> 人员管理系统信息修改 </h2>
  <table class="table table-bordered table-striped" *ngIf="hero" style="margin-top: 10px">
    <tr>
      <th> 编号 :</th>
      <td>{{hero.id}}</td>
    </tr>
    <tr>
      <th> 姓名 : </th>
      <td>{{hero.name}}</td>
    </tr>
    <tr>
```

```
      <th> 部门 : </th>
      <td><input [(ngModel)]="hero.department" placeholder="department"
                  style="border-radius: 5px"/></td>
    </tr>
    <tr>
      <th> 职务 : </th>
      <td> <input [(ngModel)]="hero.post" placeholder="post"
                  style="border-radius: 5px" /></td>
    </tr>
    <tr>
      <th> 入职时间 : </th>
      <td>  <input [(ngModel)]="hero.Entrytime" placeholder="Entrytime"
                  style="border-radius: 5px"/></td>
    </tr>
    <tr>
      <th> 电话 : </th>
      <td>  <input [(ngModel)]="hero.Telephone" placeholder="Telephone"
                  style="border-radius: 5px" /></td>
    </tr>
    <tr>
      <th> 人员性质 : </th>
      <td> <input [(ngModel)]="hero.situation" placeholder="situation"
                  style="border-radius: 5px"/></td>
    </tr>
  </table>
  <div >
    <button (click)="goBack()" class="btn btn-primary btn-lg"> 返回 </button>
    <button (click)="save()" class="btn btn-success btn-lg"> 保存 </button>
  </div>
</div>
```

在 hero-detail.component.ts 文件中，编写 save() 和 goBack() 方法，在 save() 方法中使用 hero 服务的 update() 方法来更新对人员信息的修改，然后导航回前视图。代码如 CORE0812 所示。

代码 CORE0812：hero-detail.component.ts

```
import { Component }  from '@angular/core';
import { Location }   from '@angular/common';
```

```
import { Hero }    from './hero';
import { HeroService } from './hero.service';

@Component({
  selector: 'hero-detail',
  templateUrl: './hero-detail.component.html',
  styleUrls: [ './hero-detail.component.css' ]
})
export class HeroDetailComponent {
  hero: Hero;

  constructor(
    private heroService: HeroService,
    private location: Location
  ) {}

  // 保存信息
  save(): void {
    this.heroService.update(this.hero)
      .then(() => this.goBack());
  }
// 返回到上一页
  goBack(): void {
    this.location.back();
  }
}
```

在 hero2.service.ts 更新数据。使用 HTTP 的 put() 方法来把更新的数据持久化到服务端。代码如 CORE0813 所示。效果如图 8.9 所示。

代码 CORE0813：hero2.service.ts

```
private headers = new Headers({'Content-Type': 'application/json'});
private heroesUrl = 'api/heroes2';

update(hero: Hero): Promise<Hero> {
  const url = '${this.heroesUrl}/${hero.id}';
  return this.http
```

```
// 通过一个编码在 URL 中的 id 来告诉服务器应该更新哪个英雄。
.put(url, JSON.stringify(hero), {headers: this.headers})
      .toPromise()
      .then(() => hero)
      .catch(this.handleError);
}
```

图 8.9　权限详情组件

至此，智慧工厂权限管理模块制作完成。

本项目通过对智慧工厂权限管理模块的学习，对权限管理获取数据方法具有初步了解，掌握权限管理增、删、改功能的实现方法。具有解决数据更改时出现问题的能力，为数据管理打下良好的基础。

deploy　部署
modules　单元
server　服务器
files　文件

developer　开发者
Tools　工具
Command　命令
Jasmine　测试框架
Karma　测试运行器

一、选择题

1.（　）文件夹包含着在浏览器中运行应用时所需的更多代码。

（A）index.html　（B）node_module　（C）systemjs.config.js　（D）app

2.（　）是“快速上手”仓库中安装的默认开发服务器，它被预先配置为回退到 index.html。

（A）Lite-Server　（B）Webpack-Dev-Server　（C）Apache　（D）Firebase

3. 以下哪个（　）工具和技术不是用来编写和运行 Angular 测试程序。

（A）Jasmine　（B）Karma　（C）npm　（D）Protractor

4.Angular 依赖注入的令牌可能有不同的类型。 通常是（　），也可能是字符串。

（A）类　（B）方法　（C）数组　（D）String

5. Angular 和其他新依赖需要添加到（　）中。

（A）node_modules（B）src　（C）package.json　（D）systemjs.config.js

二、填空题

1. 部署应用最简化的方式是直接把它发布到开发环境之外的 _______ 上。

2. 如果准备把该应用放在子目录下，需编辑 _______，并适当设置 <base href>。

3. _______ 在 .htaccess 文件中添加一个重写规则。

4. 独立单元测试用于测试那些完全不依赖 Angular 或不需要注入值的类实例。 测试程序会 _______，为构造函数参数提供所需的测试替身，然后测试该实例的 API 接口。

5. 使用 _______ 命令从命令行中编译并在 Karma 中运行上面的测试程序。

三、上机题

通过本项目所学的技能，把 Angular 项目部署到服务器上。